GEOTECHNICAL
GROUND
INVESTIGATION

GEOTECHNICAL GROUND INVESTIGATION

Myint Win Bo
Bo & Associates Inc., Canada

World Scientific

NEW JERSEY · LONDON · SINGAPORE · BEIJING · SHANGHAI · HONG KONG · TAIPEI · CHENNAI · TOKYO

Published by

World Scientific Publishing Co. Pte. Ltd.

5 Toh Tuck Link, Singapore 596224

USA office: 27 Warren Street, Suite 401-402, Hackensack, NJ 07601

UK office: 57 Shelton Street, Covent Garden, London WC2H 9HE

Library of Congress Cataloging-in-Publication Data
Names: Bo, Myint Win, 1954–   author.
Title: Geotechnical ground investigation / Myint Win Bo, Bo & Associates Inc., Canada.
Description: Singapore ; Hackensack, NJ ; London : World Scientific, 2022. |
    Includes bibliographical references and index.
Identifiers: LCCN 2021057691 | ISBN 9789811236723 (hardcover) |
    ISBN 9789811236730 (ebook for institutions) | ISBN 9789811236747 (ebook for individuals)
Subjects: LCSH: Soil mechanics. | Hydrogeology. | Geotechnical engineering.
Classification: LCC TA710 .B5983 2022 | DDC 624.1/5136--dc23/eng/20220107
LC record available at https://lccn.loc.gov/2021057691

British Library Cataloguing-in-Publication Data
A catalogue record for this book is available from the British Library.

For any available supplementary material, please visit
https://www.worldscientific.com/worldscibooks/10.1142/12270#t=suppl

Desk Editors: Balamurugan Rajendran/Amanda Yun

Typeset by Stallion Press
Email: enquiries@stallionpress.com

Dedicated to my wife Win Myint Than

Preface

The success of geotechnical design and construction depends on a systematically planned ground investigation. Only a properly planned ground investigation and geotechnical testing, including laboratory and *in-situ* tests, can reveal sub-surface conditions such as ground and groundwater profiles, assess geotechnical characteristics of soils and rocks, and provide representative design parameters for geotechnical analyses. The reliability of geotechnical information obtained from ground investigation can affect construction costs and the safety of the developed infrastructures considerably. Geotechnical investigation is normally carried out in the very first phase of an infrastructure development. Although ground investigations are normally planned by technical staff or engineers at the intermediate level, most of the field work is usually covered by professions at the entry level. As such, not all entry-level professionals have gone through adequate training in geotechnical investigation in their undergraduate studies. Furthermore, there are few geotechnical or ground investigation textbooks available for engineering students, early-career geotechnical engineers, or engineering geologists. In particular, there has been lack of relevant books published in recent years.

Over the past few decades, new ground investigation techniques and new or improved *in-situ* testing methods have been developed. There is a need for a new book to cover the latest developments and new testing procedures. It is hoped that this book will provide an update on some of the recent state-of-the-art or state-of-the-practice techniques on geotechnical investigation including topics related to advanced *in-situ* testing, geophysical techniques, and geotechnical instrumentation. The objective of this book is to present some topics from preliminary desk studies to intrusive ground investigations that are required for a geotechnical investigation program ranging from the planning to the reporting stages that are not commonly discussed in a textbook. The book also provides examples of data presentation and reporting formats.

This book has been written by a practicing geotechnical engineer with an academic background and extensive research experience who was deeply involved in geotechnical investigations and *in-situ* tests in many mega geotechnical investigation projects around the world spanning four continents. This book is intended to be a good reference for beginners in geotechnical investigation and geotechnical engineers who would like to upgrade their systematic geotechnical knowledge related to both planning and execution of ground investigation.

The author would like to thank Professor Victor Choa, former Dean of Students, Nanyang Technological University, Singapore, for his guidance and training given during the implementation of the extensive ground investigation work for the Changi East Reclamation Project in Singapore as well as during the author's research work leading to his Ph.D. degree at Nanyang Technological University. The author's gratitude also goes to the late Dr. A. Vijaratnam, former Chairman of SPECS Consultants Pte Ltd and former Pro-Chancellor of the Nanyang Technological University, for his mentorship, particularly during the author's services at SPECS Consultants Pte Ltd (a wholly owned subsidy of PSA Corporation), Singapore.

The author is indebted to Mr. Thet Oo, Senior Geophysicist at EGS, Hong Kong, Mr. Bernie Villegas, former Senior Associate, DST Consulting Engineer Inc., Thunder Bay, Ontario, Canada, Mr. Mike Neumann, President of Planmac Engineering Inc, Toronto, Ontario, Canada, Mr. Wesley Saunder, Director (Business Development), GKM Consultants, Ontario, Canada, Mr. Jim Greger of Geokon, USA, Mr. Rick Spillers of CME Equipment, USA, Mr. Daniëlle Span of GeoMil Equipment, The Netherlands, Mr. Diego Marchetti of Marchetti Dilatometer, Rome, Italy, and Mr. Choon Park, Principal Geophysicist, Park Seismic, USA. He would also like to extend his appreciation to Dr. Jian Chu of Nanyang Technological University who contributed the 3-D modeling section in Chapter 2 and his research collaborators Mr. Kiefer Chaim from the Building and Construction Authority, Singapore, and Mr. Kok Pang Lam from JTC Corporation, as well as his students and researchers, in particular, Dr. Shifan Wu, Dr. Xiaohua Pan, Dr. Xiaohui Qi, and Mr. Zarli Aung for their support.

Last but not least, he would like to acknowledge Mr. Karin Maheswaran of Bo & Associates Inc., for his assistance in the preparation of the manuscript for this book, and thank his family, especially his wife Win Myint Than, for their unwearied support.

About the Author

Myint Win BO is the President and Chief Executive Officer at Bo & Associates Inc., Canada. He graduated with a B.Sc. degree (Geology) from the University of Rangoon and received a Postgraduate Diploma in Hydrogeology from University College London, UK, and an M.Sc. degree from the University of London, UK. He obtained his Ph.D. in Civil Engineering (specialized in geotechnics) from the Nanyang Technological University, Singapore, and obtained a Certificate of Executive Management and Leadership from the Massachusetts Institute of Technology, USA. He is a Fellow of the Geological Society, London, UK, and a Fellow of the Institution of Civil Engineers, UK. He is also a professional engineer, professional geoscientist, International Professional Engineer (UK), Chartered Geologist, Chartered Scientist, Chartered Engineer, Chartered Environmentalist, Chartered Manager, European Geologist, and European Engineer. Dr. Bo has been serving many professional societies as a committee member at both national and international levels. Dr. Bo has worked on four continents — North America, Europe, Asia, and Oceania. Dr. Bo is an experienced practicing engineer, geologist, environmentalist, scientist, and entrepreneur as well as an educator, and he has given more than 40 special/keynote lectures and workshops at international conferences, tertiary institutions, and professional associations. Dr. Bo is also an Adjunct Professor at York University and Ryerson University, Canada, as well as at Swinburne University of Technology, Australia. He has published four textbooks, five book chapters, and over 200 technical papers for international journals and conferences. Dr. Bo is an Editor for five international journals including the *Environmental Geotechnics Journal* and

Geotechnical Research Journal published by the Institution of Civil Engineers (ICE), UK. His work has been cited more than 4700 times in referred journals and proceedings. Dr. Bo is a winner of more than $ one million in research grants and also the winner of many awards for personal achievements as well as having designed many award-winning projects around the world.

Contents

Preface   vii

About the Author   ix

List of Symbols and Abbreviations   xiii

Chapter 1   Introduction   1

Chapter 2   Preliminary Study   7

Chapter 3   Planning for Intrusive Ground Investigation   35

Chapter 4   Method of Intrusive Ground Investigation and Procedures   53

Chapter 5   Specialized *In-situ* Tests and Performance Verification Tests   89

Chapter 6   Application of Geophysical Techniques in Geotechnical
             Ground Investigation   127

Chapter 7   Sampling   153

Chapter 8   Geotechnical Instrumentation   183

Chapter 9   Data Management and Reporting   209

References   231

Index   243

List of Symbols and Abbreviations

P_0	absolute initial system pressure
P_t	absolute pressure at time t
K_a	active earth pressure
θ	angle from vertical line
θ_1, θ_2	angle normal to the boundary made by current lines
i	angle of ray
c'	apparent cohesion
T	basic time factor
E_m	blow count correction factor
C_b	blow count correction factor
Cr	blow count correction factor
C_N	blow count correction factor
N_F	blow count correction factor
k	bulk modulus
C	coefficient of consolidation or swelling
C_v	coefficient of consolidation
C_{v0}	coefficient of consolidation
C_h	coefficient of consolidation due to horizontal flow
C_c	compression index
v_p	compression velocity
P	compression wave

Vp	compressional wave velocity
N_k	cone factor
CPT	cone penetration test
CPMT	cone pressuremeter test
q_c	cone resistance
C_{vr}	consolidation in recompression range
K	constant depending upon dimensions and shape of the vane
H_c	constant head
M	constrained modulus
q_t	corrected cone resistance
N_{60}	corrected SPT blow count
N_{cor}	corrected SPT blow count
A	cross sectional area of borehole, casing, or standpipe
i_c	critical angle
A	cross sectional area
$C1, C2$	current electrode
I	current in amperes
I	current in conducting body
$Z-$	depth
Z	depth of boundary
D_b	depth of boring
d	diameter of filter (mm)
D	diameter of intake section
E_D	dilatometer modulus
DMT	dilatometer test
N'	dilation correction
T	dimensionless time factor
T_{fllex}	dimensionless time factor
X	distance between source and geophone
Φ	drained friction angle
ϕ'	drained friction angle
φ	drained strength parameters

K_0	earth pressure at rest
σ'_{va}	effective overburden stress
E	elastic modulus
u_e	equilibrium pore pressure measured from CPTU test
k_h	horizontal hydraulic conductivity
K_D	horizontal stress index
k	hydraulic conductivity
I	influence value
V_0	initial gas volume (mL)
σ_0	initial stress
u_i	initial pore pressure
ID	inner diameter
F	intake factor
τ	lag time
λ	Lame's compressibility
L	length
L	length measured along casing
L	length of filter (mm)
L	length of intake section
L	length of test section (m)
PL	limit pressure
I_D	material index
D	measured width of the vane
H	measured height of the vane
P_0	membrane lift off pressure
MAM	micro-tremor array measurements
MASW	multi-channel analysis of surface waves
NC	normally consolidated condition
S	number of storey
m	numerical Variable
OD	outer diameter
N_{cor}	overburden correction

δ_v	overburden stress
OCR	overconsolidation ratio
K_p	passive earth pressure
k	permeability of soil
Ip	plasticity index
σ	Poisson's ratio
V	potential difference between two surfaces of constant potential
E	potential difference
$P1, P2$	potential electrode
U_{bt}	pore pressure at cone base
u_t	pore pressure at time t
n	porosity
u_0	pre-insertion pore water pressure
P_1	pressure required for center of membrane to deflect by a pre-set distance of 1 mm into soil
H	pressure head of water in the test section (m)
P	primary seismic wave
ρ	radius of cavity
R	radius of pushing cone (m)
r	radius of a sphere equal in surface area to that of the cylindrical tip
r	radius of test section (m)
q	rate of flow
Q	rate of inject in cubic meter per sec (m³/s)
C_r	recompression index
ρ	resistivity
ρ_1, ρ_2	resistivities of formation 1 & 2
R	resistance in ohms
S	secondary seismic wave
C_α	secondary compression index
SCPT	seismic cone test
SDMT	seismic dilatometer test
SBPT	self-boring pressuremeter test

L	separated distance between two equipotential surfaces
F	shape factor
μ	shear modulus
S	shear wave
v_s	shear wave velocity
f_s	sleeve friction
n	slope of the q, $1/\sqrt{t}$ graph
SP	spontaneous potential
N	SPT blow count
U_0	static pore water pressure
M_0	strain shear modulus
$\Delta\sigma_z$	stress increase at depth Z
σ_z	stress at depth Z,
q_∞	steady state of flow
q_0	surface or contact stress
t	time elapsed to reach a given degree of consolidation in years
T_{50}	time factor
t_{50}	time taken for excess pore pressure to fall half of its maximum
M	torque to shear the soil in $N\,m$
τ_f	torque measured undrained shear strength
t	transmission time
s_u	undrained shear strength
C_u	undrained shear strength
a	unequal area ratio
V_1	velocity along upper formation
B	width
E	Young's modulus

Chapter 1

Introduction

The term "ground" is defined in BS 5930 (1999) as covering soils, rocks, and made ground (e.g., reclaimed land). While Field Investigation refers to investigating the whole site and surrounding field for all aspects, Site Investigation investigates only the whole site for all aspects. However, **Ground Investigation** only investigates the ground and below ground aspects of the site, which may include soils and rocks, groundwater, and contaminants. This book will cover the scope of the geotechnical aspect of ground investigation, covering soils, rocks, and groundwater with special emphasis on soil.

Ground Investigation is required for foundations and ground engineering works, which are usually included in infrastructure developments. In order to obtain the ground profile, groundwater conditions, and geotechnical parameters, ground investigation must be carried out at the proposed infrastructure development site. During the ground investigation, *in-situ* testing is necessary to obtain soil geotechnical parameters and collections of suitable types of disturbed and undisturbed samples for visual inspection and further laboratory testing such as classification tests and strength and consolidation tests. These collected disturbed and undisturbed samples are tested in an accredited geotechnical laboratory to obtain the required geotechnical parameters. The geotechnical data and parameters obtained from the properly planned geotechnical investigation can only provide valid analyses and design output. In order for the geotechnical engineer to plan an appropriate ground investigation which could provide all the necessary information, advanced knowledge of the site and surrounding area is required. To understand the site and the surrounding area, one has to start with a desk study, which means searching the information available in the public domain, literature, and data collections.

1.1 Desk Study

Desk study normally involves collecting information on the site condition, such as its topography, drainage, access condition, site geology, groundwater condition, as well as hydrogeological information, any hazard and risk related to the ground and development, any record of historical features which are necessary to maintain, and so on. Most of the information described above could probably be obtained from government institutions such as the Geological Survey Department, mining department, archeological department, and aerial and land survey department. Aerial photography, and existing geological, hydrogeological, and geotechnical reports of the site and surrounding area can also be good references. In many cases, the clients themselves may have reports and records related to their previous works, which could be valuable information to start with. As old, abandoned mines always pose a risk of foundation failure, records and maps of abandoned sites, their conditions, and closure records are also useful information as part of risk assessment and management of the foundation and underground structure developments. During the desk study stage or in the planning stage, available utility service location maps and archeological maps are usually reviewed in order to be able to plan for any drilling, excavation, and penetration testing to be sufficiently distanced from those locations so as to prevent damaging them. The details of a desk study are extensively discussed in Chapter 2 of this book.

1.2 Reconnaissance Survey

In many cases, large-scale and large-area infrastructure projects, such as mine infrastructure developments, highways and railways, dams and reservoirs, and land reclamation, may require reconnaissance surveys such as a drive-over or flyover survey, aerial photography survey, Light Detection and Ranging (LiDAR) survey, and geophysical survey to collect preliminary data. These surveys are normally planned and carried out based on the information obtained from the desk study. These surveys could reveal the site topography, drainage, and relief information as well as preliminary large-scale ground profile and rock and soil types. The details of these types of reconnaissance surveys are described in Chapters 2 and 6.

1.3 Intrusive Ground Investigation

Intrusive ground investigation is planned based on the information obtained during the desk study and through a reconnaissance survey, if any. At the planning stage, based on the type of infrastructure, the complexity of the proposed infrastructure, ground and groundwater conditions of the site, the level of impact to the environment, and risk, an appropriately suitable category of ground investigation is adopted. The higher the degree of complexity and the risk, the more comprehensive and higher the category of the ground investigation planned. In some cases, simple forms of ground investigation such as test pitting and trenching are used for simple low-rise structures with low-risk development on uniform competent ground conditions. These types of simple excavations to reveal the sub-surface ground conditions are possible with hand excavation or machine excavation using an excavator, depending on the type of ground being excavated and the depth of excavation. The possible depths of excavation used in test pitting and trenching are usually limited to a shallow depth of 3 to 4 m. Both disturbed and undisturbed samples such as block sampling are possible with this method. Observation of groundwater seepage and measurement of groundwater level are possible within the excavation.

Most geotechnical investigations to explore greater depths are commonly carried out by advancing investigation boreholes using a suitable type of drilling rig applying the appropriate method of drilling. Only then can the necessary ground profiling, collections of samples, and *in-situ* testing be carried out. The number and positions of boreholes are selected based on the type of infrastructure project the investigation is for. Factors such as area, linear or vertical projects, their size, dimensions and geometry, complexity of ground and groundwater conditions, and potential geotechnical issues also affect this. The depths of investigation are also determined based on the type of foundation infrastructure and their potential influence depth, depth to the groundwater level, depth to the competent formation and extent of poor ground condition, etc. The method of drilling adopted is selected based on the expected formation to be penetrated and potential issues of advancing the borehole, such as the relative density and stiffness of soils, content of gravels and boulders and quality of rocks and groundwater conditions. In addition to the method of drilling, the type of mounting for the drilling equipment should also be considered based on the conditions of access and terrain. Various methods of drilling, drilling techniques, and types of mounting are described in Chapter 4.

1.4 Sample Collection

During the borehole drilling investigation, the representative soil samples are collected for visual classification purposes and further geotechnical laboratory testing. Depending on the type of soil encountered during the drilling, suitable samples are collected. A simple rule of thumb determines that granular soils, which easily disintegrate due to a lack of cohesion and cementation, are usually collected as disturbed samples. Disturbed samples could be recovered using the split spoon sampler used for the Standard Penetration test. Alternatively, disturbed samples can be collected from auger returns or by collecting suspended solids in the drilling fluid returns. Undisturbed samples can be collected from soil types with a reasonable cohesion and can be obtained using specially designed samplers which introduce minimal disturbance to the soil. While disturbed samples are only suitable for visual inspection and geotechnical classification tests, undisturbed samples are suitable for carrying out strength and consolidation tests. Samples are required to be stored, transported, and extruded via suitable methods which minimize post-sampling disturbance. The details in the methods of sampling and available samplers, possible sample disturbances during sample collection, and methods of storage, transportation, and sample extrusions are described in Chapters 3 and 7.

1.5 *In-situ* Testing

To observe resistances of each type of soil and pore water conditions, the drilling process is commonly accompanied by applying forces and pressures using specific *in-situ* testing equipment and methods. Various types of commonly used *in-situ* tests, such as the Standard Penetration Test (SPT), Field Vane Shear Test (FVT), and a few different types of hydraulic conductivity tests, are described in the chapter on intrusive ground investigation (Chapter 4). In addition to the commonly used *in-situ* tests, there are specialized *in-situ* tests, such as the Cone Penetrometer (CPT), Flat Dilatometer (DMT), Cone Pressuremeter (CPMT), Seismic Cone Test (SCPT), Seismic Dilatometer (SDMT), Dynamic Cone, and BAT Permeameter, which can be carried out without advancing the borehole. These are described in detail in Chapter 5. These specialized *in-situ* tests allow the classification of soils from the measured parameters without the need to advance the borehole or collect a sample. In addition, direct measured parameters can be processed and interpreted to obtain geotechnical parameters such as strength, modulus, compressibility and stiffness, coefficient of consolidation, and stress history. Details of specialized *in-situ* testing and methods of interpretation are extensively discussed in Chapter 5.

1.6 Geophysical Survey

In addition to the geotechnical *in-situ* testing methods, both surface geophysics and sub-surface geophysical logging are also applied when profiling sub-surface ground and groundwater conditions. Seismic surveys and resistivity surveys are commonly used in the ground profiling of large-scale project sites as a preliminary survey to assist in planning an intrusive ground investigation. Various types of probes, such as the Resistivity probe, Gamma Probe, Neutron probe, and Spontaneous Potential probe, and downhole and crosshole seismic testing have been used in geotechnical ground investigation to profile the ground or measure soil parameters such as modulus, moisture content, and shear wave velocity. Details on the application of geophysical techniques in geotechnical ground investigation are described in Chapter 6.

1.7 Geotechnical Instrumentation

In many cases, some basic geotechnical instruments, such as a groundwater monitoring well or gas well, are usually installed to obtain the baseline conditions of a site's groundwater, gas and to collect groundwater and gas samples if any exist. These wells are also maintained for a certain duration for further monitoring purposes. In addition to the commonly used geotechnical instrumentation regularly accompanying geotechnical investigation, there are some geotechnical instruments which are required to be installed during the geotechnical ground investigation to collect baseline information. Such instruments are kept for long-term use during as well as soon after construction. These instruments can monitor the effects of construction activities on neighboring infrastructure and the surrounding environment. These instruments include the settlement plate and point, piezometer, water stand point, inclinometer, and deep reference point. Details regarding the use of these instruments, their installation, how they are used for monitoring and analyses, as well as typical presentation formats are described in detail in Chapter 8.

1.8 Factual Reporting

After carrying out any geotechnical investigation involving drilling, sampling, *in-situ* testing, installation of geotechnical instruments, monitoring, data processing, laboratory testing, and data analyses and interpretation, the field records and processed data are required to be presented in the form of a factual report—in other words, as a geotechnical baseline report or geotechnical information report. Chapter 9 describes the methods of field description and identification, classification of soil, and the preparation of bore logs and their presentation. In addition, the presentation of the soil profile in the form of cross sections, a fence diagram, and a ground model is also described. Chapter 9 describes how to prepare factual reports outlining the method of investigation, presentation of laboratory data, and interpretation of essential geotechnical parameters and their presentations.

This book is intended to be a useful basic handbook and guide for beginners in geotechnical engineering practice as well as for practicing geotechnical engineers and engineering geologists.

Chapter 2

Preliminary Study

Preliminary studies are generally carried out as a desk study in an office and a site visit in the form of a walkover, drive-over, or flyover depending upon the size of the area to be covered. Based on the information collected from the desk study and site visit, a reconnaissance survey may be carried out before moving onto intrusive ground investigation.

2.1 Desk Study

The desk study generally collects all related information available in the public domain, such as websites associated with institutional and regional authorities and literature stored and available through a database and library. Some geotechnical investigation and engineering companies in the industry usually have their own database for the information related to the projects in which they were involved in the past within and around the areas of interest. The literature includes published papers available through peer review journals, conference proceedings, and post-graduate theses, in addition to the project reports.

Reviewing the following available maps for information is useful while performing the desk study before any intrusive survey is conducted:

- Topographic maps
- Digital and satellite maps
- Google maps
- Historical land use maps
- Geological maps
- Hydrogeological maps
- Drainage maps
- Archeological maps and

- Maps showing abandoned mines and mine shafts
- Utility maps

Topographic maps show both horizontal distance and vertical distance by means of elevations with reference to certain datum. Topographic maps showing spot levels and/or showing contour levels are useful to study the ground variation and surface features. These types of site-specific topographic maps could be available from previous projects carried out at the area of interest and in many cases are commercially available to purchase from state-owned survey departments in many countries. Various kinds of scales varying between 1:10,000 and 1:1,000,000 are available, with 1:50,000 scales being the most commonly available scale, among others. Examples of such maps are those published by the Federal Government of Canada ranging between 1:50,000 and 1:250,000 scales and those published by the Geological Survey in the United States ranging between 1:24,000 and 1:250,000 scales. These commercially available topographic maps also show water features such as streams, springs, lakes, swamps, ponds, and wetlands, and major man-made features such as roads, railways, power lines, telecommunication lines and pipelines, airports, seaports, and cities as well as natural features like trees.

Alternatively, digital and satellite maps provide good visual features of the ground in both 2-Dimensional and 3-Dimensional views. Nowadays, digital maps are available online in many countries. Figures 2.1–2.3 show typical topographic maps with contours, relief, and satellite images, respectively. Nowadays, maps produced by Google are extensively used in engineering aspects.

In addition to the commercially available topographic survey maps, site-specific topographic surveys from previous projects could also be available from databases maintained by local and regional authorities, public institutions, and private consulting companies. Figures 2.4 and 2.5 show an existing spot elevation map and a topographic elevation contour map, respectively, from one of the previous projects.

Records of historical land use maps are useful in determining past cut and fill exercises as well as the existence of potential old mines and adits. In many countries, records of abandoned mines and mine shafts are available from government agencies. Figure 2.6 shows historical land use of the area.

Regional and local geological maps are usually available from the national and provincial/state geological survey departments together with brief or detailed geological survey reports. Two types of geological maps useful for study are superficial maps and bedrock maps. A superficial map usually describes shallow drift deposit such as quaternary unconsolidated geological formations, whereas a bedrock map describes types of bedrock formations with succession based on geological age. Formations are again subdivided into series, systems, and groups. In bedrock maps, structural geology such as dips and strikes of bedrocks and their structural features such as faults, folding, bedding, jointing, foliations, and cleavages are also

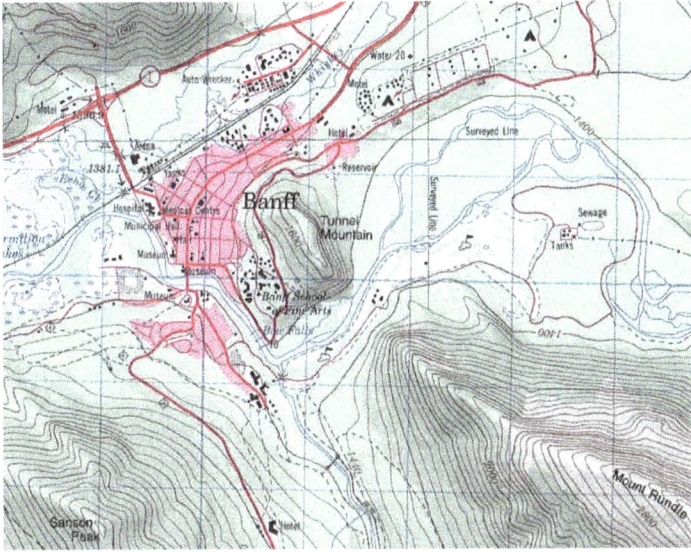

Figure 2.1   Topographic map showing variation of elevations with contours.
Source: www.canmaps.com.

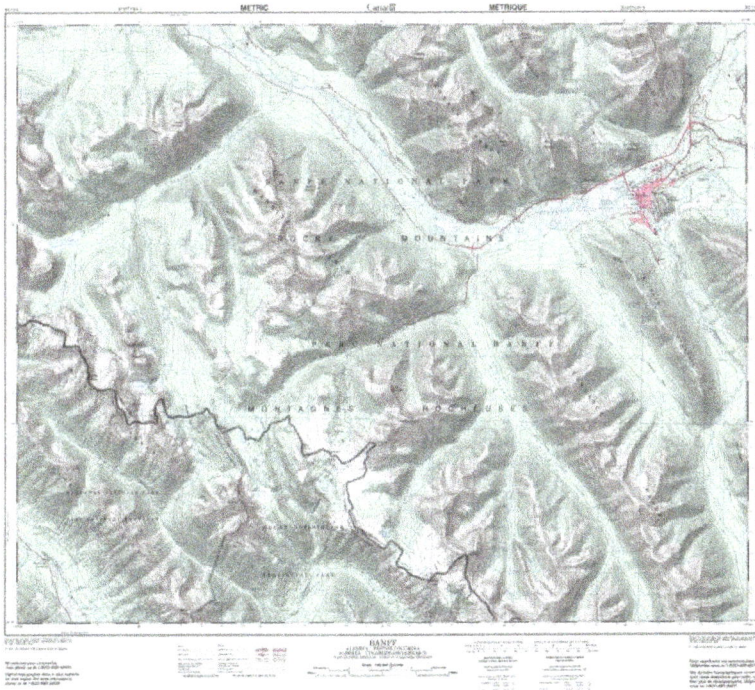

Figure 2.2   Map showing relief surrounding Banff, Alberta, Canada.
Source: www.canmaps.com.

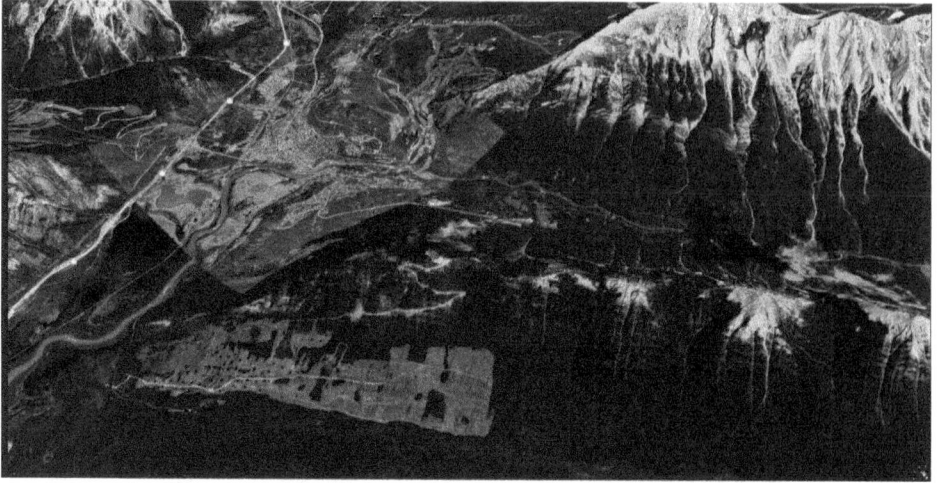

Figure 2.3   Satellite image from Google earth.

Source: Google © 2021 Maxar Technologies.

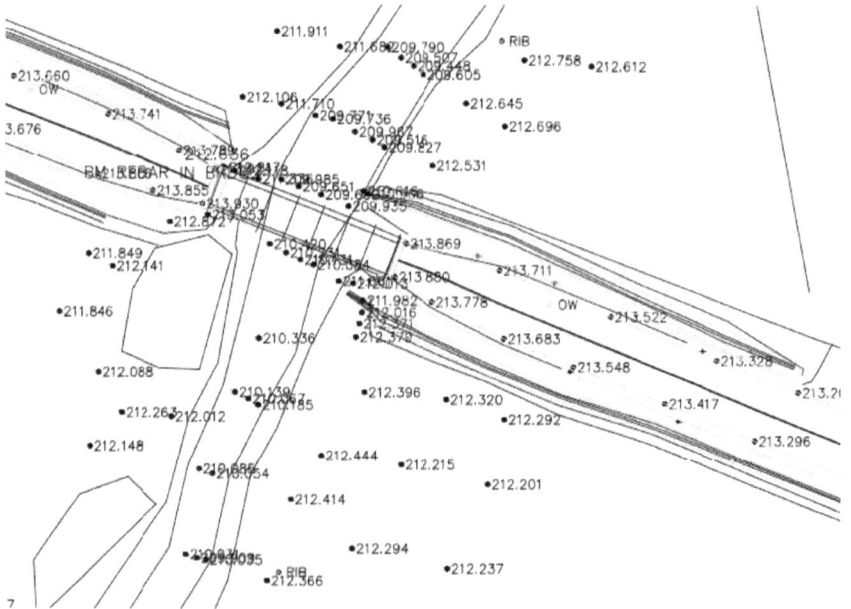

Figure 2.4   Survey map showing spot levels from previous project.

Source: Courtesy of Planmac Engineering Inc.

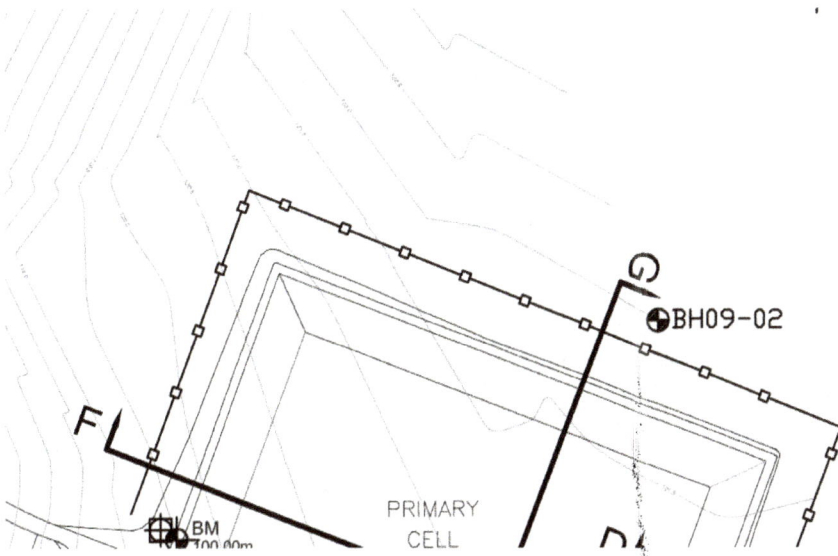

Figure 2.5   Survey map showing elevations with contour from previous project.
Source: Courtesy of Bo and Associates Inc.

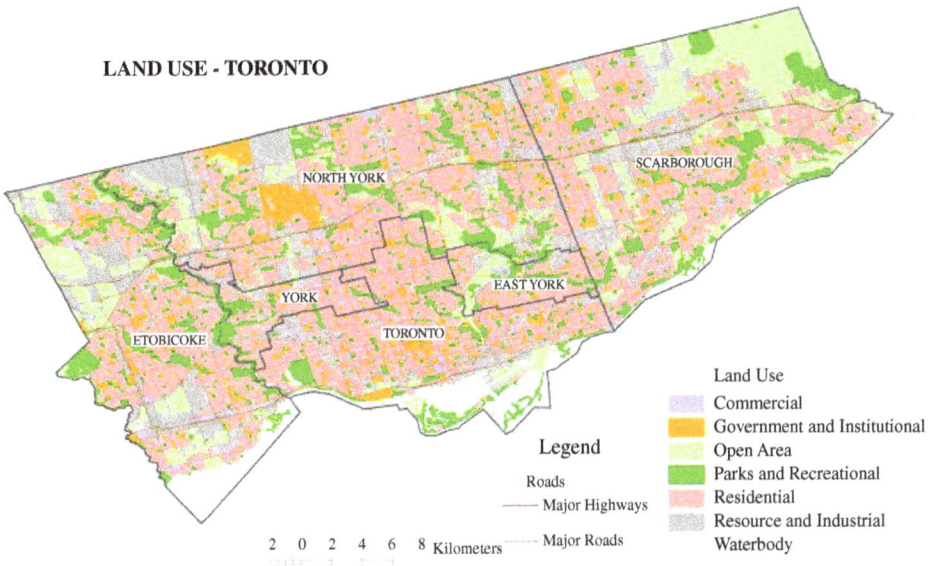

Figure 2.6   Toronto land use map.
Source: DMTI CanMap Route Logistic.

described. Figures 2.7 and 2.8 show examples of a superficial map and a bedrock geology map, respectively. In some countries and some regional areas, information such as depth to the bottom of soft deposits, depth to bedrock, and cross-sectional geological profiles is available. Figures 2.9 and 2.10 show a cross-sectional profile of bedrock geology and the depth to bedrock cross section, respectively. Figure 2.11 shows the thickness of soft deposits from a large area project and Figure 2.12 shows the depth to competent formation from a land reclamation project.

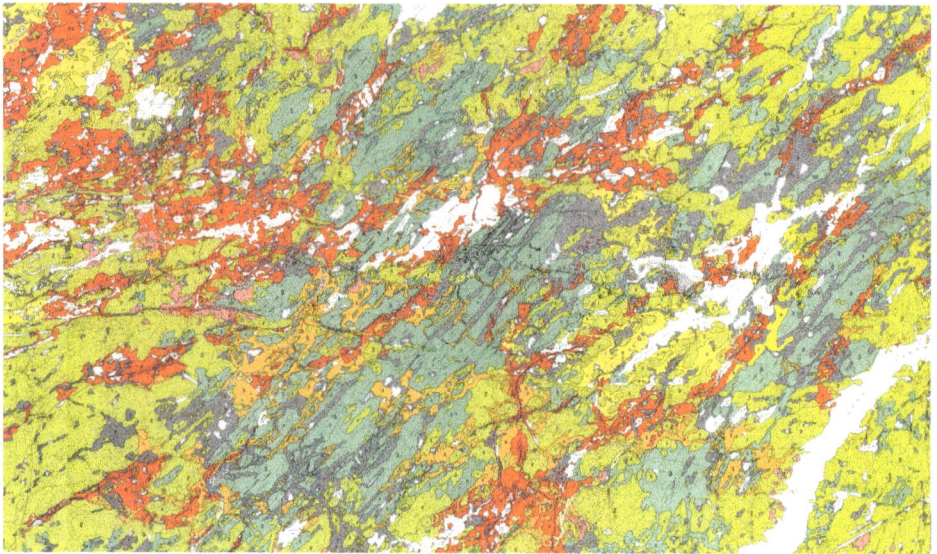

Figure 2.7   Example of superficial map.

Source: Reproduced with permission of the Department of Natural Resources, 2021. © Queen's Printer for Ontario, 1991. Kristjansson and Thorleifson (1991).

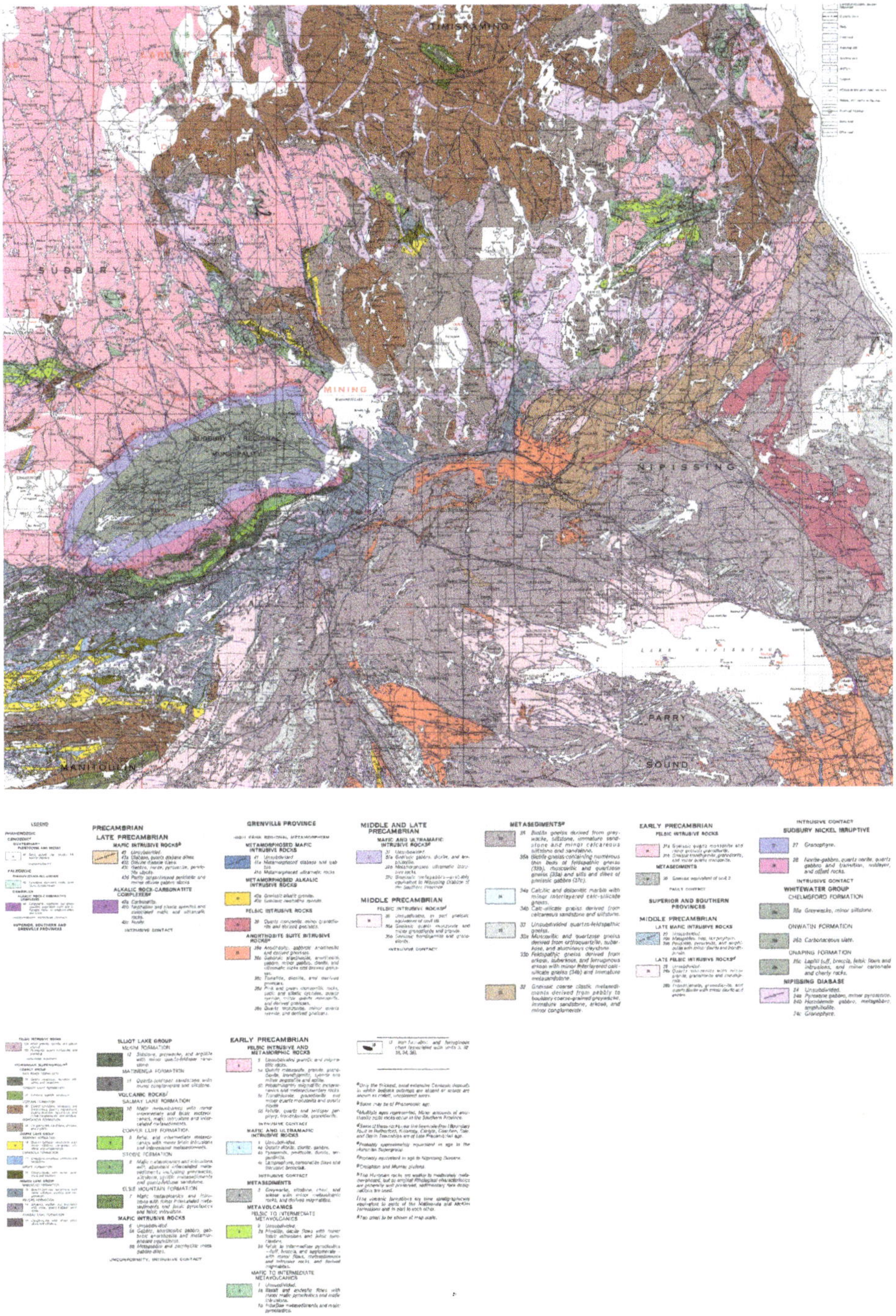

Figure 2.8   Example of bedrock geology.

Source: Reproduced with permission from Queen's Printer for Ontario, 1977.

Scales of geological maps that vary between 1:24,000 and 1:2,500,000 are available in many countries. Depending upon the scale of the maps, the features of geological information provided vary. It could also vary depending upon the publishing institutions and countries.

Regional and local hydrogeological maps are available from the national and provincial/state departments of geological survey or water resources departments. Hydrogeological maps will provide groundwater tables and piezometric levels as well as information on unconfined aquifers, confined aquifers, and related potential hydraulic pressure and hydraulic properties of aquifers, such as hydraulic

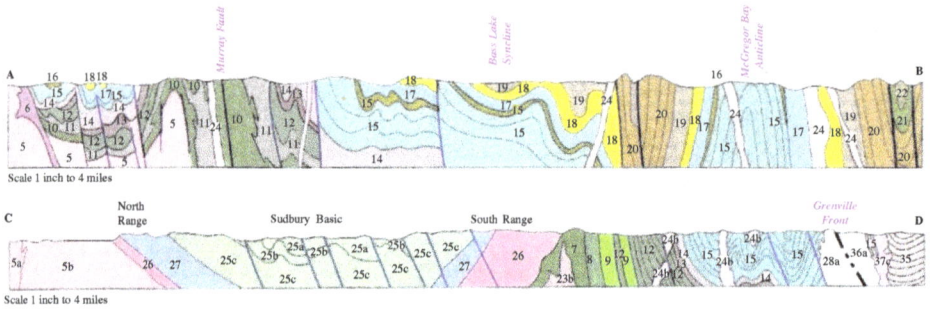

Figure 2.9   Cross-sectional showing bedrock geology.

Source: Reproduced with permission from Queen's Printer for Ontario, 1977.

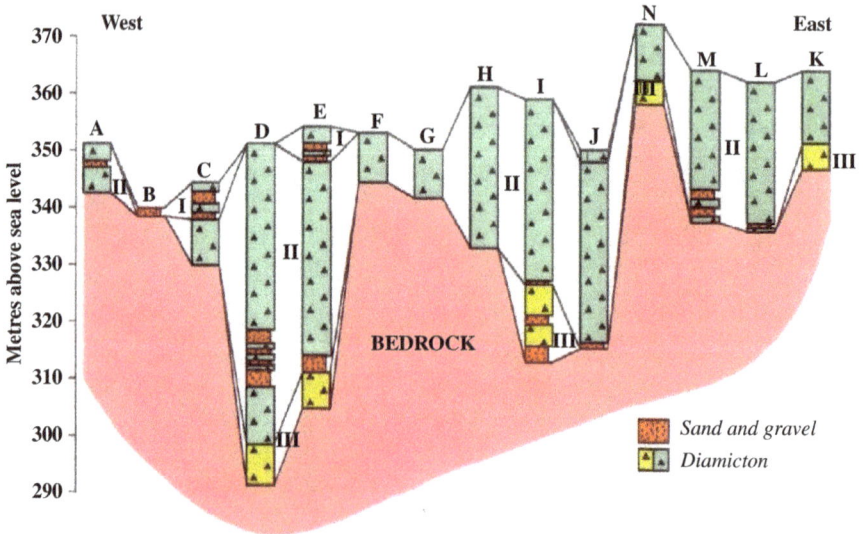

Figure 2.10   Depth to bedrock cross section.

Source: Reproduced with permission of the Department of Natural Resources, 2021. © Queen's Printer for Ontario, 1991.

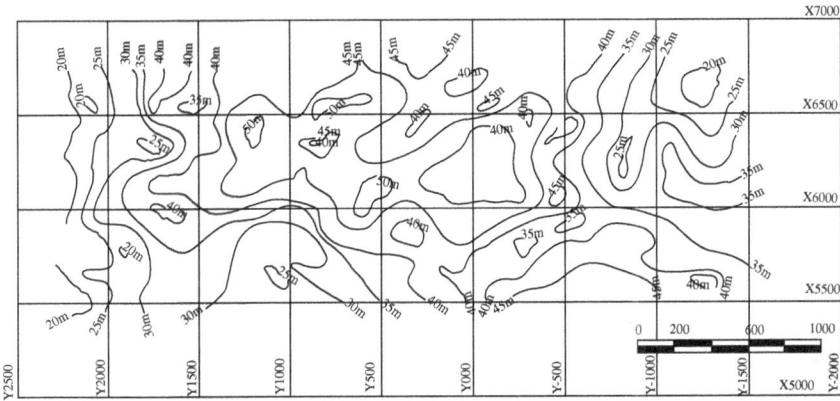

Figure 2.11   Contour showing thickness of soft deposits.

Source: Bo *et al.* (2003).

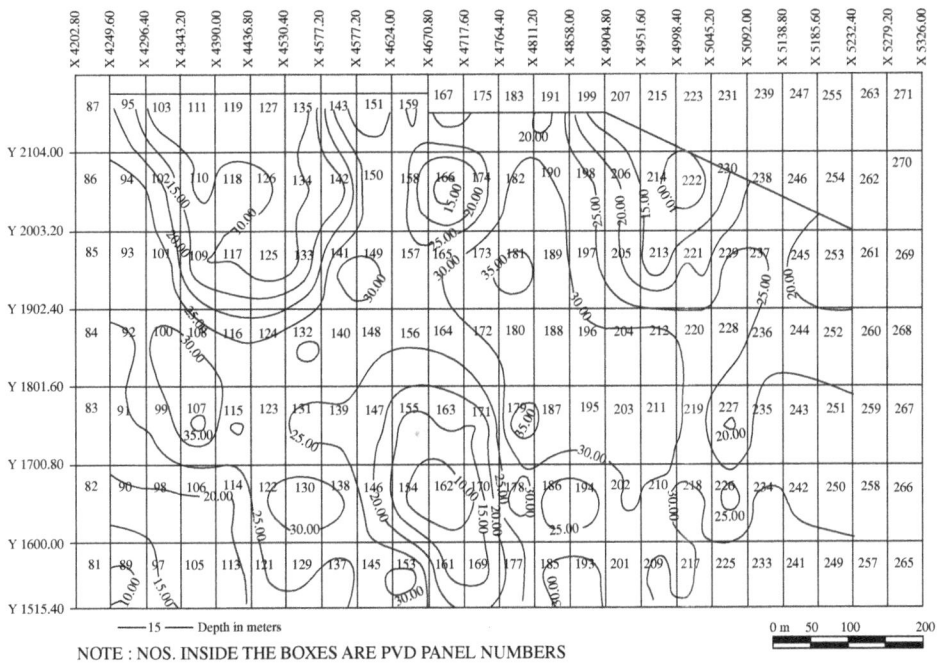

NOTE : NOS. INSIDE THE BOXES ARE PVD PANEL NUMBERS

Figure 2.12   Contour showing depth to competent formation.

Source: Bo *et al.* (2003).

conductivity, hydraulic gradient, and storage coefficient. In some countries and some regional areas, information such as depth to groundwater table, elevation of ground-water table, and piezometric elevations, as well as depth to aquifer and cross-sectional profiles of aquifers and groundwater profiles, is available. Some examples of

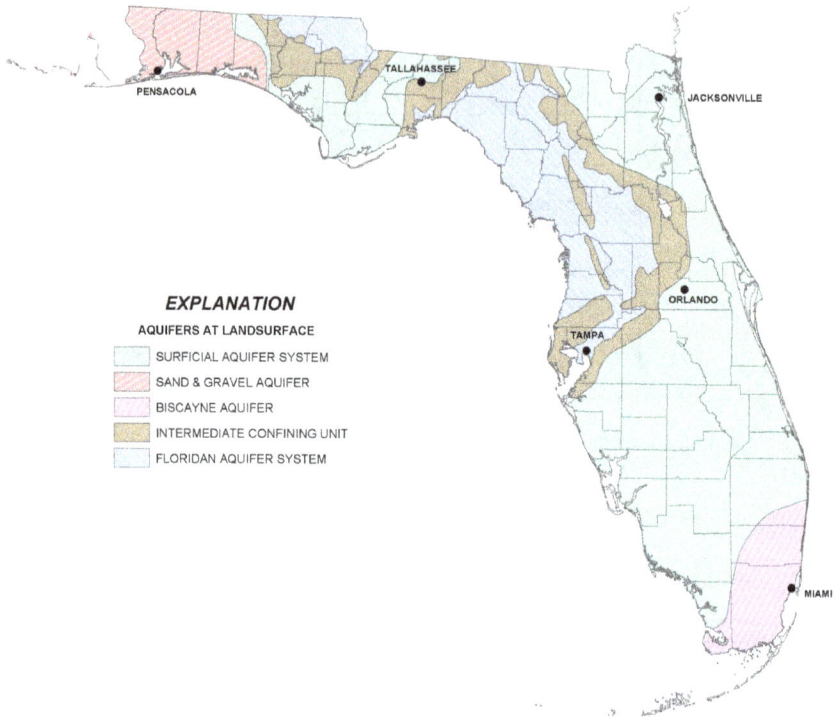

Figure 2.13   Aquifer system map of the state of Florida.
Source: Courtesy of U.S. Geological Survey.

those are many hydrogeological maps available from the State of Florida, USA. Figures 2.13 and 2.14 show the aquifer system map and a cross-sectional diagram of the aquifer system of Florida State, USA, respectively. Figure 2.15 shows contours of thickness of the aquifers in the Florida state, USA. Figures 2.16 and 2.17 show the potentiometric elevation contours of groundwater head and transmissivity distributions across the State of Florida, USA, respectively.

In Canada, the following types of maps are commercially available through the Go Trekkers website for online purchase, which is certified by Natural Resources Canada (NRCan) for Canadian topographic map printing:

- Topographic Maps
- Digital Topographic Maps
- Canadian Hydrographic Nautical Charts/Maps
- Geological Survey Maps published by Geological Survey of Canada

As an example, in the province of Ontario, Canada, the Ministry of Energy, Northern Development and Mines has a good database that consists of the

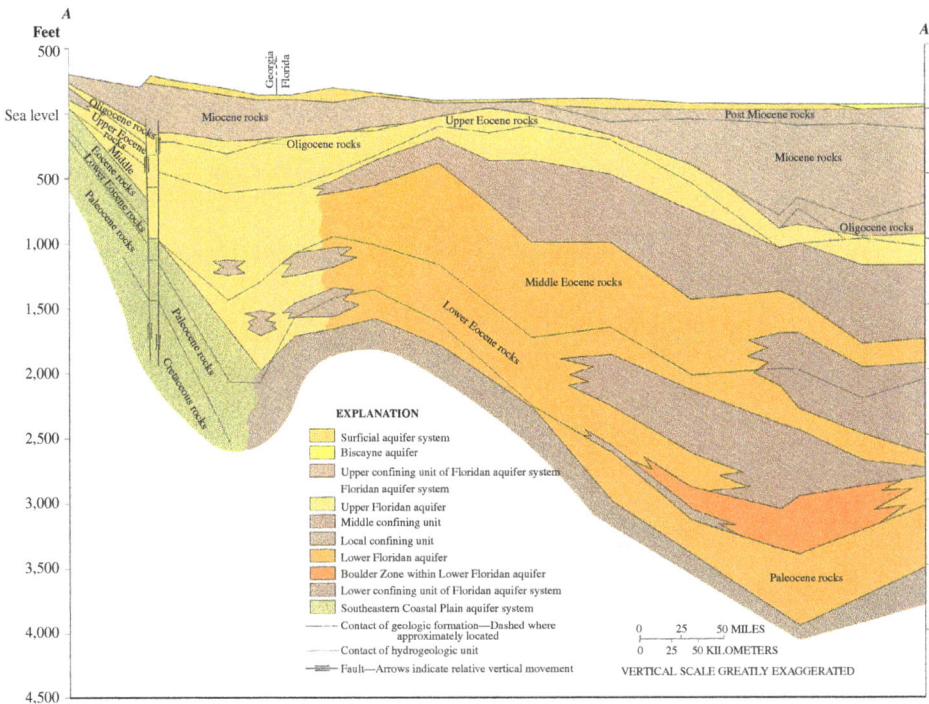

Figure 2.14　Cross-sectional diagram showing aquifer system in the state of Florida.
Source: Courtesy of U.S. Geological Survey.

following maps for Ontario published by Ontario Geological Survey, which are available online:

- Surficial Maps consisting of the following information:
  - Quaternary Geology
  - Drift Thickness
  - Bedrock Topography
  - Engineering Geology
  - Physiography
  - Geotechnical Studies
- Bedrock Geology Maps consisting of the following information:
  - Precambrian
  - Paleozoic
  - Geochronology
- Geophysical Survey Maps consisting of the following:
  - Magnetic Survey
  - Electromagnetic and Electrical Surveys

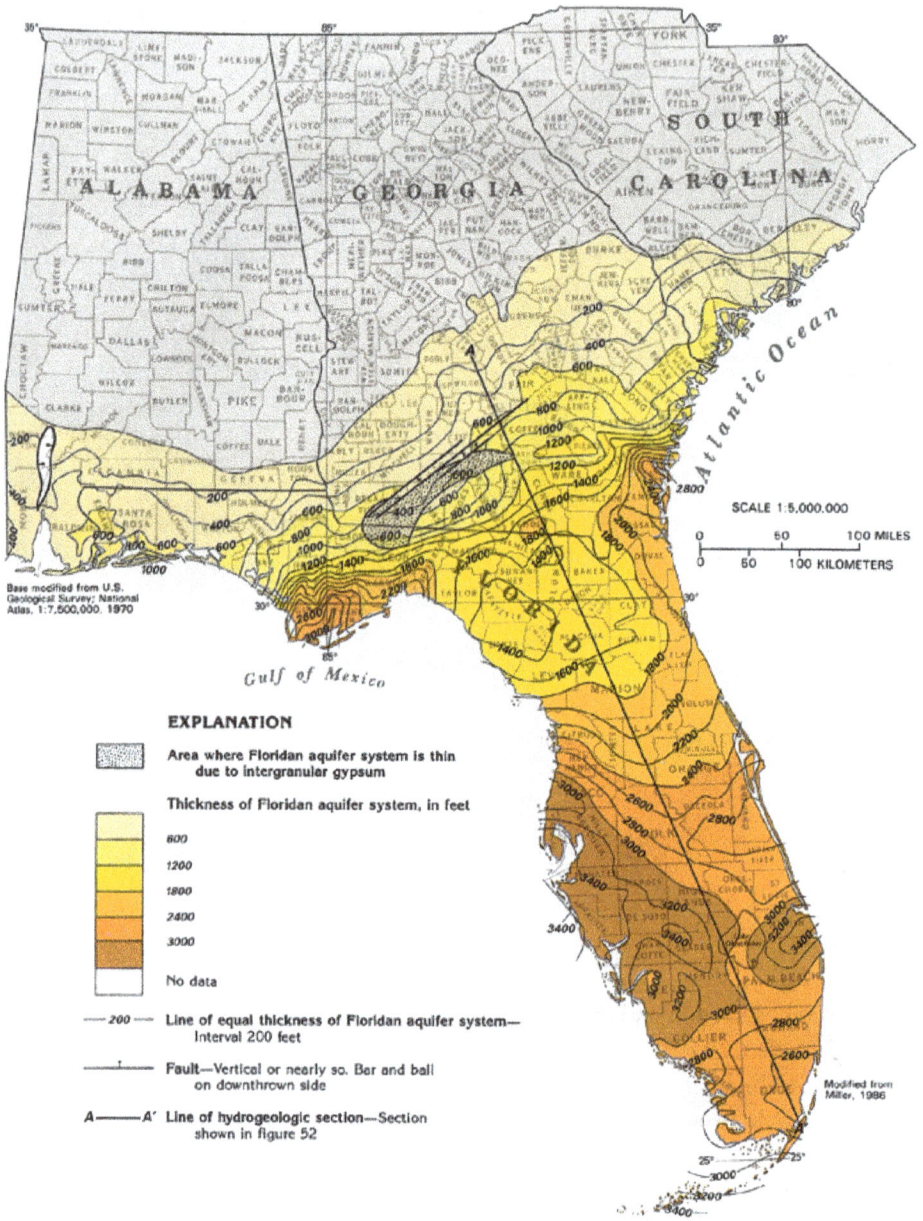

Figure 2.15   Contour showing aquifer thicknesses in the state of Florida.

Source: Courtesy of U.S. Geological Survey.

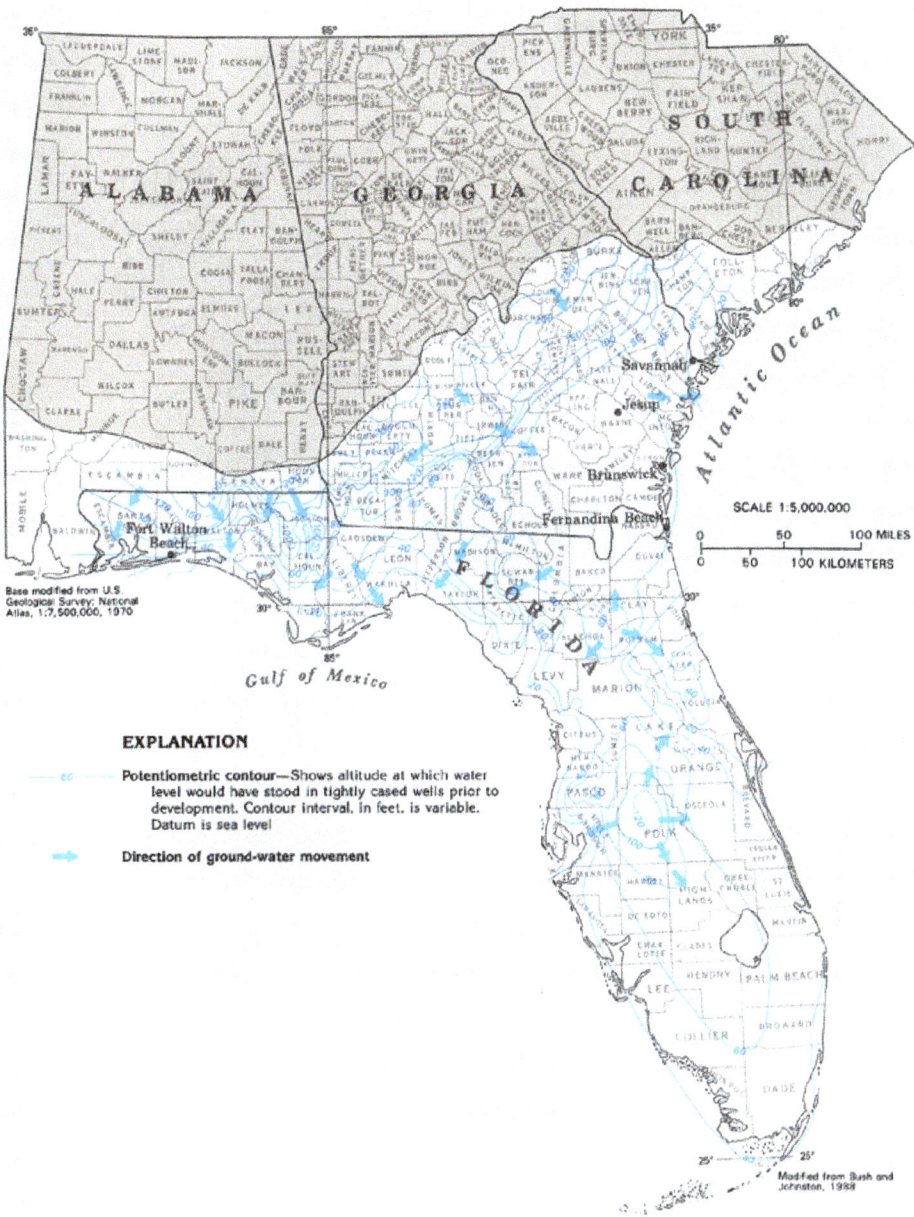

Figure 2.16   Contour showing potentiometric elevation of groundwater levels in the state of Florida.
Source: Courtesy of U.S. Geological Survey.

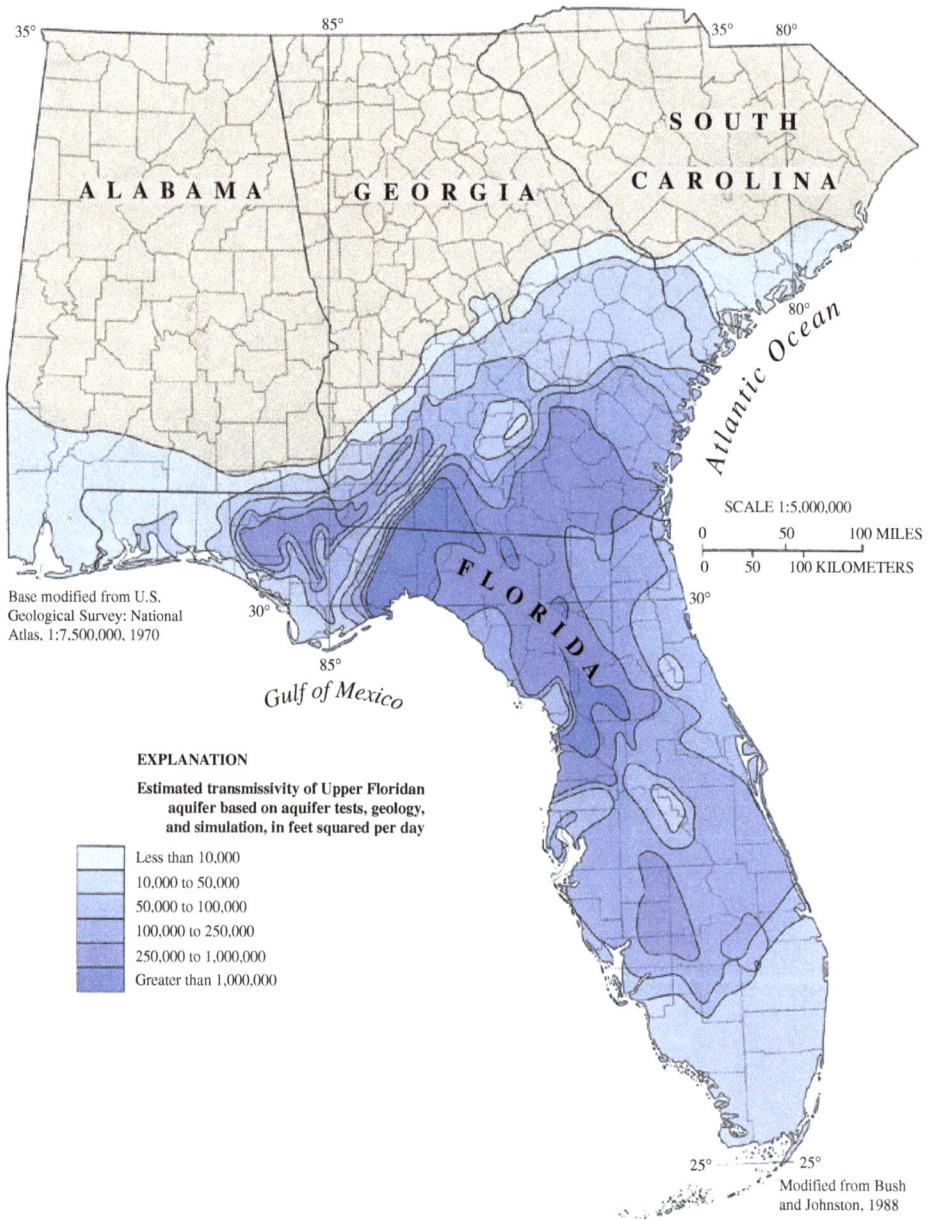

Figure 2.17   Map showing estimated transmissivity of groundwater levels in the state of Florida.
Source: Courtesy of U.S. Geological Survey.

○ Spectrometer Surveys
○ Gravity and Seismic Surveys
- Geochemical Survey Maps consisting of the following information:
  ○ Water and Sediment Geochemistry
  ○ Overburden Geochemistry
  ○ Bedrock Geochemistry
  ○ Soil Geochemistry
- Abandoned Mine Database which is available to view on Google Interactive Maps

Figure 2.18 shows drift thickness of the area and Figure 2.19 shows abandoned mine sites.

In addition to these maps, other information which is useful before intrusive surveys is as follows:

- Regional and Local Geology
- Regional Hydrogeology
- Borehole database
- Well records database
- Any abandoned mine records
- Any past geotechnical and/or hydrogeological investigation reports for infrastructure developments

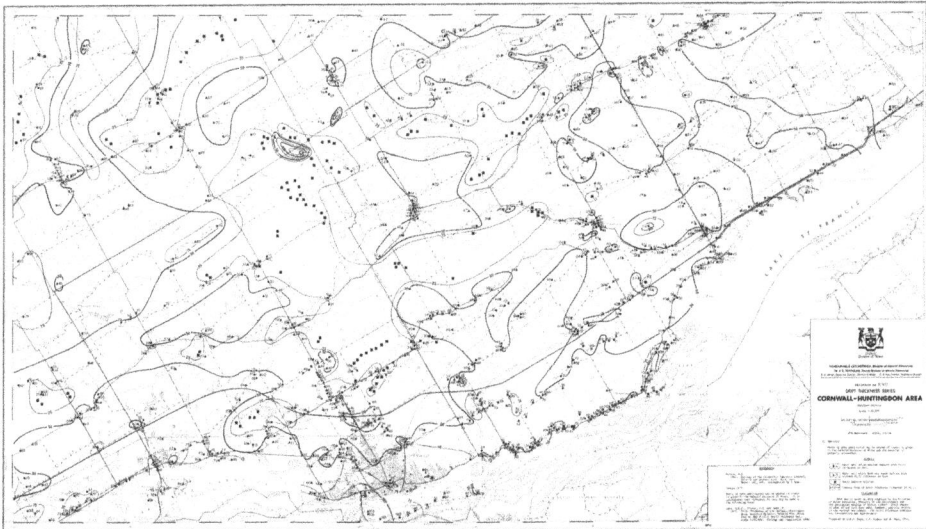

Figure 2.18   Drift thickness contour map.

Source: Reproduced with permission from © Queen's Printer for Ontario, 1975. Gwyn *et al.* (1975).

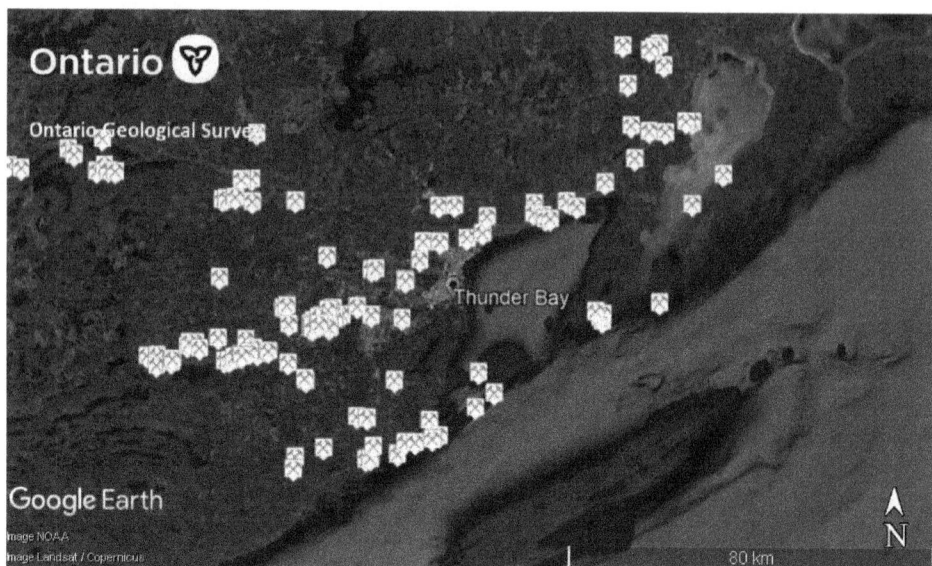

Figure 2.19   Map showing abandoned mine sites.
Source: Reproduced with permission from © Queen's Printer for Ontario, 1975.

- Any available geotechnical characterization technical papers and reports of the areas
- Any available construction records such as excavation, piling, embankment construction, and tunneling.

As an example, in the province of Ontario, Canada, the Ministry of Energy, Northern Development and Mines has a good collection of databases consisting of geological survey reports, geophysical survey reports, abandoned mine reports for Ontario published by Ontario Geological Survey, and miscellaneous papers published from 1891 till the present, which are available online. The Toronto Regional Conservation Authority of Canada and the Ontario Ministry of the Environment, Conservation and Parks have massive records of groundwater monitoring data for the province of Ontario. Well records are also available through interactive maps from the Ontario Government, Canada (Figure 2.20). Geotechnical borehole datasets are also available to the public through the online Ontario Data Catalogue. Tables 2.1 and 2.2 show examples of well records and geotechnical borehole data, respectively, at the study area.

Some of the example references of notable literature in Canada are as follows:

- Urban Geology of Canadian Cities by Karrow and White (1988)
- The Hydrogeology of Southern Ontario by Singer, Cheng and Scafe (1997) published by the Ministry of Environment and Energy

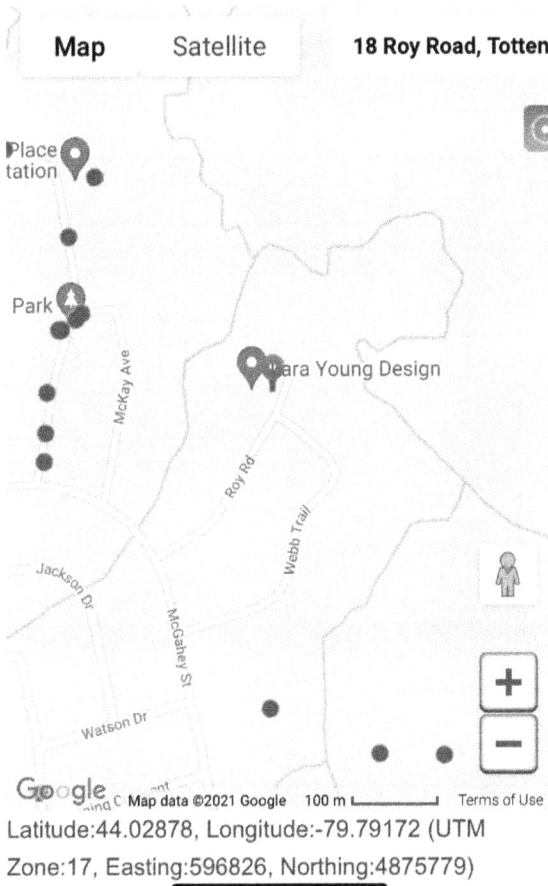

Figure 2.20   Existing groundwater wells in the study area.

Source: ©2021 Google. Iteractive Map from Ontario Government website, CANADA.

- Canadian Geotechnical Journals published by the Canadian Geotechnical Society
- Proceedings of Annual Conferences of the Canadian Geotechnical Society

2.2 Site Reconnaissance Survey

Site reconnaissance surveys are carried out in different ways depending upon the size and magnitude of the projects. A small size such as residential, commercial, or institutional infrastructure developments could be covered with a walkover survey.

During the walkover survey, topography, drainage, vegetation, any water features, overhead service lines and above-ground services, existing infrastructure, access conditions, and any other special features could be visualized and recorded.

Table 2.1   Well records at the study area.

Bo & Associates Inc.			Existing Water Well Data Surrounding Study Area					
2233 Argentia Road, Suite 302			Project ID:			GS-20212030		
Mississauga, ON			Project Location:			18 Roy Road, Tottenham, Ontario		
L5N 2X7			Proximity:			500 meter radius		
Canada								*available from public domain*
Sr Number	Well ID	Date Installed (mm/dd/yyyy)	Location	Well Depth (m)	Static Water Level (m)	Screen Depth (m)	Type of Aquifer Formation	Remarks
1	7271212	09/01/2016	NW	-	-	-	-	Abandoned
2	7271211	09/01/2016	NW	-	-	-	-	Abandoned
3	7271210	09/01/2016	NW	-	-	-	-	Abandoned
4	7279467	08/19/2016	NW	-	-	-	-	-
5	7271209	09/01/2016	NW	-	-	-	-	Abandoned
6	5732355	07/16/1996	W	24.7	24.7	23.2-24.7	CSND	Domestic
7	5738726	04/23/2004	W	15.5	13,8.53,14.5	-	SILT, SAND, CLAY	Domestic
8	7271208	09/01/2016	W	-				Abandoned
9	7271207	09/01/2016	W	-				Abandoned
10	7259057	11/18/2015	W	-	-	-	-	-
11	7250001	07/27/2015	SW	-	-	-	-	-
12	7355736	07/07/2019	SW	9.8	4.57	8.22-9.75	SILT, FSND, CLAY	Monitoring
13	7234143	11/20/2014	SW	-	-	-	-	-
14	5719870	08/18/1984	SW	10.4	3.05	-	SAND	Domestic
15	5704129	09/14/1965	S	19.2	17.1	-	CLAY,STNS	Domestic
16	5709430	10/12/1972	S	18.9	18.2	-	SAND	Domestic
17	5712728	07/28/1975	S	15.2	13.7	-	SAND	Domestic
								Bo & Associates Inc.

Source: Courtesy of Bo and Associates Inc.

Large area projects covering a large area or linear projects covering long distances such as highways, transmission lines, and pipelines can only be covered by flyover or drive-over surveys. While flyover surveys can cover brief information collections such as preliminary reconnaissance surveys and collect similar information like that described in the walkover surveys, they can also collect additional information such as potential alignment, water crossings, depressions and ranges, potential water crossings, cut and fill or tunnel routes, potential crossings of swamps, and rock outcrops.

Flyover surveys are normally carried out for extremely large area projects such as mining projects, land reclamation projects, and the abovementioned long stretch projects such as linear projects. Similar information as described above is collected for such projects.

Other large-scale reconnaissance surveys are aerial photography, LiDAR surveys, and geophysical surveys, which will be described in the subsequent sections.

2.3 Aerial Photography

Aerial photographs are taken from an aeroplane. These photographs can be taken rapidly at the speed of the aeroplane, without worrying about access conditions.

Table 2.2   Existing geotechnical borehole records at the study area.

Bo & Associates Inc. 2233 Argentia Road, Suite 302 Mississauga, ON L5N 2X7 Canada				Existing Geotechnical Borehole Data Surrounding Study Area			
				Project ID:		GS-20211001	
				Project Location:		2233 Argentia Road, Mississauga, Ontario	
				Proximity:		500 meter radius	
							available from public domain
ID	Location from Site	Approxi. Distance	Date Completed	Depth (m)	Water Level (m)	Layer Depth (m)	Layer Description
641817	S	120	1970	9.8	0.4	0-0.5	soil, silt
						0.5-4.3	till, silt, clay, gravel, brown hard
						4.3-5.5	till, shale brown, dense
						5.5-9.8	bedrock, shale, red, dense
641815	S	215	1970	8.7	0.2	0-0.6	till, silt, clay
						0.6-2.1	till, shale, boulders, compact
						2.1-4.7	bedrock, shale, red, dense
						4.7-8.7	bedrock, shale, red
641814	SE	300	1970	6.1	-	0-0.3	soil, organic material, black
						0.3-2.1	till, silt, clay, sand, brown, compact
						2.1-2.4	silt, sand, grey, dense
						2.4-3	till, shale, red, dense
						3-6.1	bedrock, shale, red, dense
641813	SE	450	1970	4.7	-	0-0.3	soil, black
						0.3-3	till, silt, clay, brown, dense
						3-4.7	till, shale, dense
641821	W	110	1970	7.7	-	0-0.2	soil
						0.2-4.6	till, silt, clay, gravel, brown, hard
						4.6-5.2	till, shale, red, dense
						5.2-6.4	bedrock, shale, red, hard
						6.4-7.7	bedrock, shale, red, hard
650238	N	120	1972	4.9	-	0-0.3	soil
						0.3-4.9	till, silt, sand, gravel, brown, dense
641820	E	265	1970	3	-	0-0.4	soil
						0.4-0.9	till, sand, gravel, brown, compact
						0.9-1.6	till, silt, clay, brown, hard
						1.6-1.8	till, shale, hard
						1.8-3	bedrock, shale, red, dense
641819	E	440	1970	4	-	0-0.3	soil
						0.3-3	till, silt, clay, gravel, brown, hard
						3-4	shale, red, hard
638980	NE	380	1970	2.8	-	0-2.4	till, clay, silt, sand, brown, stiff
						2.4-2.8	till, shale, compact
650240	N	300	1912	6.1	-	0-0.3	soil
						0.3-3	till, silt, sand, gravel, grey, dense
						3-6.1	shale, limestone, red, hard
638983	N	410	1970	2.7	-	0-1.7	till, clay, silt, sand, compact
						1.7-2.7	till, shale, red, moist
650241	N	410	1971	4.3	-	0-0.2	soil
						0.2-1.5	till, silt, sand, gravel, grey
						1.5-4.3	shale, limestone, red, hard
638984	N	450	1970	6.1	-	0-2.1	till, gravel, sand, silt, brown, stiff
						2.1-3.1	till, silt, sand, gravel, dense
						3.1-5.3	sand, brown, compact
						5.3-6.1	shale, till, red, hard
650239	NW	265	1972	6.4	0.5	0-0.3	soil
						0.3-5.1	till, silt, sand, gravel, brown, dense
						5.1-6.4	clay, silt, sand, grey, hard
							Bo & Associates Inc.

Source: Courtesy of Bo and Associates Inc.

Nowadays, with the help of the Global Positioning System (GPS) and available satellites, the altitude of the flight, the height of the location, and their positions can be produced much more accurately than those produced a few decades ago. Aerial photographs are taken with many overlaps which allow one to prepare a collection of photographs of the given area to form the required information of the area. Such photographs usually have 60% overlap and are termed Mosaics. The typical scales available for aerial photography are 1:8,000, 1:20,000, 1:40,000, and 1:60,000 with larger scales providing more details. Figure 2.21 shows the concept of aerial photo taking.

The application of a stereoscope allows one to examine overlapping air photos to view the large terrain section in 3-dimensional view. Reliefs seen through these processes may be exaggerated and this should be taken into consideration during

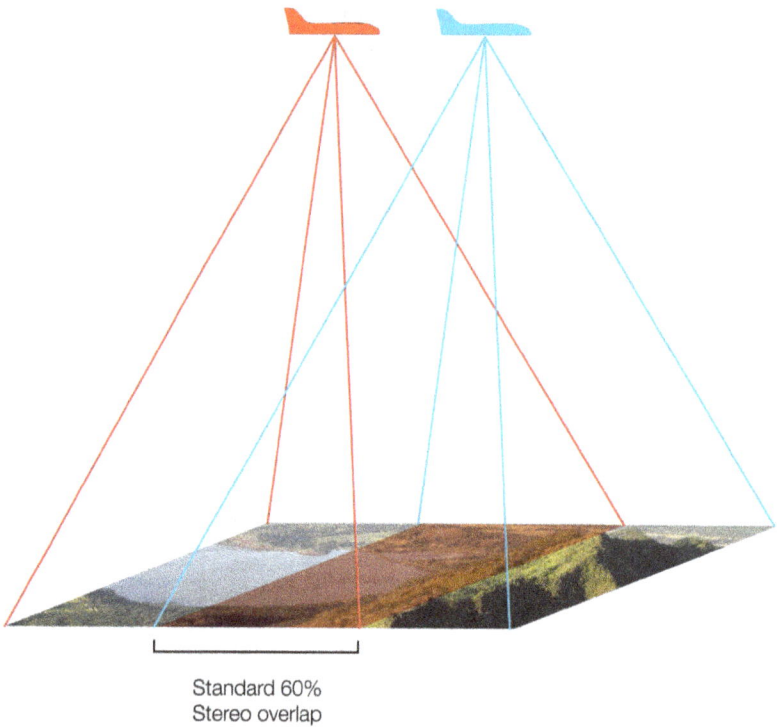

Figure 2.21   Concept of aerial photo taking.

Source: © Commonwealth of Australia — on behalf of the Intergovernmental Committee on Surveying and Mapping.

interpretation. When air photos are viewed through a stereoscope, the following objects are identified:

- Shape
- Size
- Pattern
- Shadow
- Tone
- Texture
- Relationships

Figure 2.22 shows a stereoscope being used in air photo interpretation. In the examination of the abovementioned objects, the followings features can be interpreted:

- Landform
- Drainage pattern

Figure 2.22   Stereoscope used for interpreting air photos.
Source: Courtesy of National Collection of Aerial Photography, NCAP.

- Gully shape and gradient
- Vegetation cover
- Land use

Figure 2.23 show two air photos overlapping 60%.

Among the features, examining drainage patterns can indicate the characteristics of underlying soils and rocks as these underlying materials control the way water flows through. Details in interpreting the rock type and structure based on basic drainage patterns can be found in various geomorphology and aerial photogrammetry textbooks.

2.4 Light Detection and Ranging Survey

A Light Detection and Ranging (LiDAR) survey can be carried out using an aeroplane from which light applying laser pulses are shot to the ground and reflected light from the ground is registered with sensitive detectors attached to the aircraft. The recorded data are analyzed and used to produce a topographic map ignoring trees and other ground cover. Accuracy and precision of maps produced using LiDAR are reported to be 100 mm. Figure 2.24 shows a LiDAR survey in progress and Figure 2.25 shows an example of a LiDAR Survey map.

Figure 2.23   Typical set of air photos with 60% overlap.
Source: City of Toronto Archives.

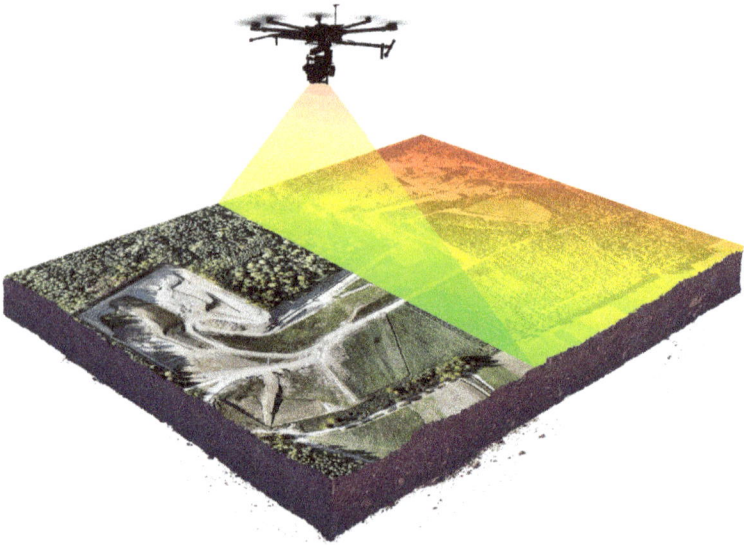

Figure 2.24   Concept showing LiDAR survey in progress.
Source: Wingtra A G.

2.5 Geophysical Survey

There are various geophysical methods which could be carried out from air, on the ground or surface, on water, and within boreholes. Airborne and seaborne surveys are usually carried out as preliminary reconnaissance surveys for large areas and the resulting profiles are verified and confirmed by exploratory boreholes. Surface geophysical methods are also carried out for large areas, but are comparatively

Figure 2.25   Map produced from LiDAR Survey.
Source: Courtesy of Airborne Imaging.

smaller than airborne survey areas. Surface geophysical surveys usually require access clearance. Geophysical testing carried out in boreholes using probes requires drilling of boreholes in the soil and rock formations in advance.

Details of the application of these methods will be described in Chapter 6.

2.6 3-D Geological Models

2.6.1 *Recent development in geo-database*

With the rapid development in computer technologies, digital data management and 3-D geological modeling have increasingly been used as alternatives to the existing 2-D geological maps and routine databases. In the UK, the British Geological Survey (BGS) has produced digital geological maps (DigMap) and attributed 3-D geological models (LithoFrame) as its primary outputs. The BGS database holds more than three million records, relating to more than 600,000 boreholes. The data are stored in standardized coded formats that ensure interoperability between BGS and other data systems (More details can be found at http://www.bgs.ac.uk/services/ngdc/management/-geology/borehole.html). The National Geotechnical Properties Database, NGPD (Self and Entwisle 2006; Self *et al.* 2012), is also maintained by the BGS Information Management (Geology)

Project. The information is supplied to BGS by clients, consultants, and contractors either as paper or pdf reports and/or as Association of Geotechnical and Geo-environmental Specialists (AGS) digital data transfer format. The design of the database is currently based on the AGS industry standard for Electronic Transfer of Geotechnical and Geo-environmental Data version 3.1.

2.6.2 *Common 3-D geo-data mapping and modeling software packages*

The most common software packages used for building 3-D geological maps and models include GOCAD, SubsurfaceViewer, Leapfrog Geo, 3-D GeoModeller, EarthVision, ArcGIS, Multilayer-GDM, GeoVisionary, Isatis, Move, Petrel, Rockworks, SKUA, and Surfer (Kessler *et al.* 2009). Of these, SubsurfaceViewer, 3-D GeoModeller, and Multilayer-GDM have been developed by geological survey organizations (GSOs) to meet customized geological mapping and modeling needs of the organizations. Many other software packages are also used in GSOs worldwide as part of modeling workflows, and these include software for GIS, geostatistical analysis, seismic depth conversion, visualization, and property modeling.

2.6.3 *Procedures of 3-D geological modeling*

3-D geological modeling has developed dramatically over the past several decades. As it is not possible to directly access the sub-surface except through digging holes, most of this understanding has to come from various indirect acquisition processes and interpreted results. In the following, the procedure for 3-D geological modeling will be explained using an example from Singapore (Pan *et al.* 2018).

Although Singapore is a small country, it geology is quite complex, which comprises granitic and sedimentary rocks, old alluvium soils, and soft marine clay. A huge amount of geological borehole data have been obtained over the years from the various constructions works carried out in Singapore. These data were used for better understanding of spatial organization of sub-surface structures of Singapore and the building of a 3-D model of the sub-soil layer.

SubsurfaceViewer was used as the 3-D geological modeling software. The methodology of 3-D geological modeling was based on a simple philosophy — the construction of geological sub-surface models has to proceed with an understanding of the complete geological sequence and the likely geomorphological evolution of the study area. Therefore, except for first preparing the geological data which could provide useful information about both the surface and underground conditions of the earth, the interpreted cross sections were created with geological knowledge to show geological relationships and depict a representation of the geological sub-surface arrangement. Then, the fence diagram was constructed using the established cross sections which can help geologists have a good understanding of the complex sub-surface arrangements of the study area. Finally, 3-D geological models were built layer by layer based on the fence diagram and geological boundaries.

2.6.4 *Geological data preparation*

Borehole data, which are direct observations of the sub-surface, are probably the most important geological data for 3-D modeling. The quality and quantity of the borehole data collected originally are of crucial importance for the accuracy of the resulting 3-D model. Data-cleaning procedures involving removing typing errors and duplicates from the database have been conducted to minimize the number of errors that affect the quality of data. The East Coast area of Singapore has been selected as an example area for 3-D geological modeling. The distribution of borehole data of this area is shown in Figure 2.26. 2-D geological maps can help to identify the extent, distribution, and borderlines of each geological

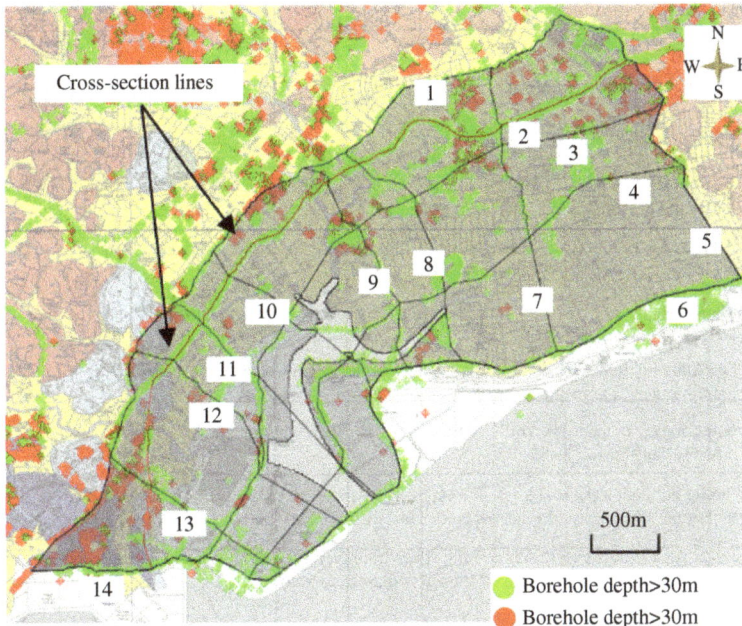

Figure 2.26   The distribution of borehole data and geological cross sections of the East Coast area (Gray color covered area) of Singapore.

Source: Pan *et al.* (2018).

formation at the earth's surface. DEMs are used to determine the top surface of the top geological formation. Archival publications for geological descriptions can share geological information of the investigated area and help geologists establish the 3-D models more efficiently. To allow all the data to be combined in the 3-D geo-modeling package, the easting and northing coordinates of the collected geological data should spatially reference with the same coordinate system.

2.6.5 *Interpreted cross section creation*

Cross sections can be created following the below procedures. First, cross-sectional lines are established based on the quality and quantity of the borehole data. As shown in Figure 2.26, fourteen cross-sectional lines have been defined along the boreholes marked in green color whose depths are greater than 30 m.

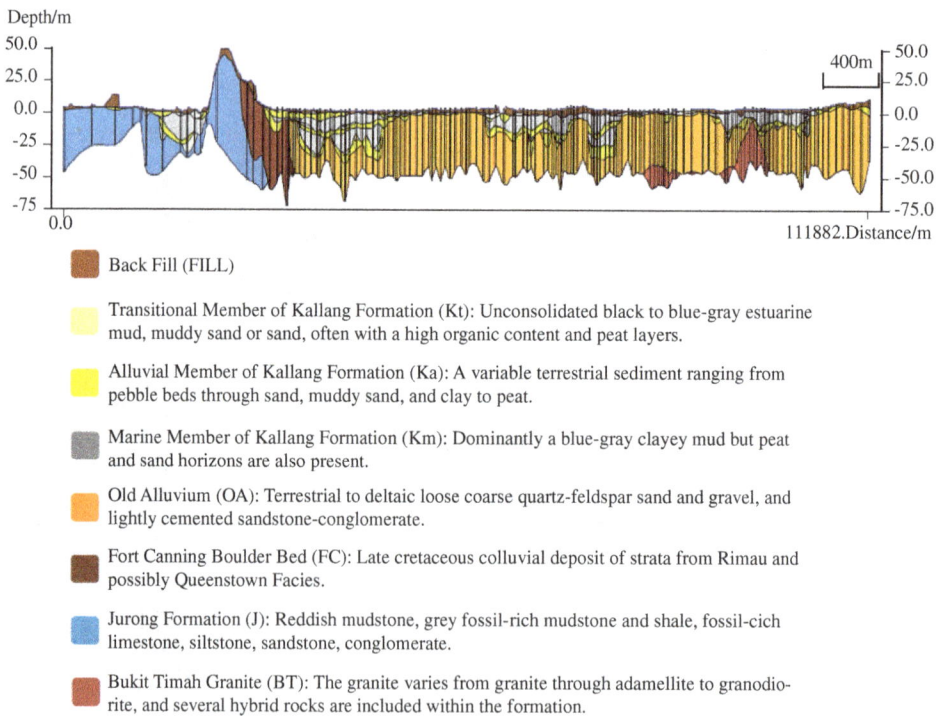

Back Fill (FILL)

Transitional Member of Kallang Formation (Kt): Unconsolidated black to blue-gray estuarine mud, muddy sand or sand, often with a high organic content and peat layers.

Alluvial Member of Kallang Formation (Ka): A variable terrestrial sediment ranging from pebble beds through sand, muddy sand, and clay to peat.

Marine Member of Kallang Formation (Km): Dominantly a blue-gray clayey mud but peat and sand horizons are also present.

Old Alluvium (OA): Terrestrial to deltaic loose coarse quartz-feldspar sand and gravel, and lightly cemented sandstone-conglomerate.

Fort Canning Boulder Bed (FC): Late cretaceous colluvial deposit of strata from Rimau and possibly Queenstown Facies.

Jurong Formation (J): Reddish mudstone, grey fossil-rich mudstone and shale, fossil-cich limestone, siltstone, sandstone, conglomerate.

Bukit Timah Granite (BT): The granite varies from granite through adamellite to granodiorite, and several hybrid rocks are included within the formation.

Figure 2.27   Geological cross section along the No. 2 cross-sectional line of the East Coast area of Singapore.

Source: Pan *et al.* (2018).

Second, one can check the integrity and validity of the borehole data along the cross-sectional lines using borehole log profiles, especially the Lithostratigraphical classification code in the "*.blg" format file which was used to interpret the Singapore formation layers. Third, high-quality boreholes along the defined cross-sectional lines are selected and presented in Section-Window, and the cross-sectional lines of different formation layers are interpreted and constructed from the top to bottom layers. The top cross-sectional line is controlled by the trace of the DEM line. The cross section along the No. 2 grid line is shown in Figure 2.27 as an example.

2.6.6 *Fence diagram construction*

In the East Coast area of Singapore, a total of fourteen interpreted cross sections were created. Using these cross sections, the fence diagrams of the East Coast area of Singapore were constructed as shown in Figure 2.28.

2.6.7 *3-D geological model building*

Based on the constructed fence diagrams, 3-D geological formation layers were built layer by layer as shown in Figure 2.29. The processes that form the geological units of each layer and their subsequent arrangement cannot be automatically

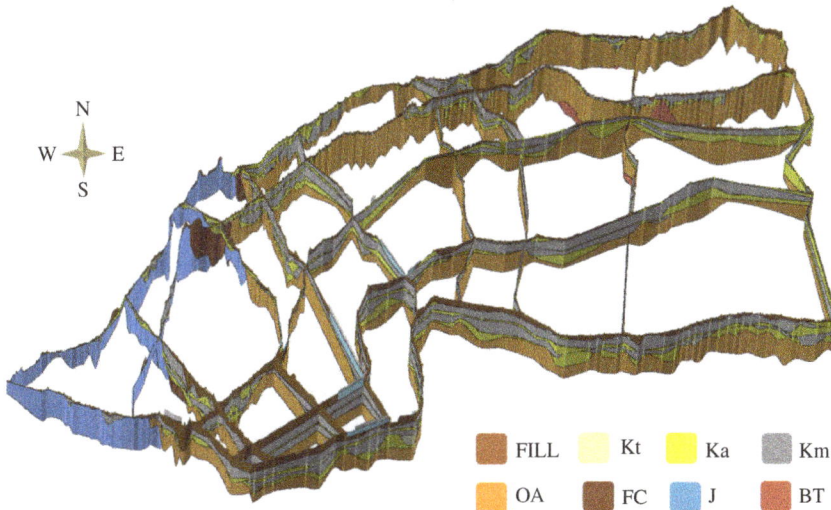

Figure 2.28   The fence diagrams of the East Coast area of Singapore.
Source: Pan *et al.* (2018).

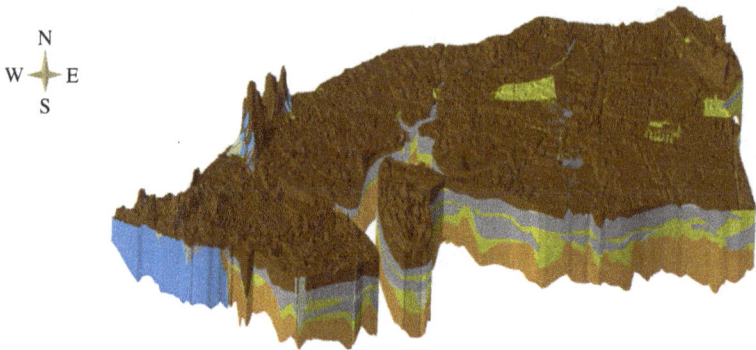

Figure 2.29   3-D geological model of the East Coast area of Singapore.
Source: Pan *et al.* (2018).

simulated accurately by software. Hence, these processes can only be captured and expressed by the sensible construction of each of the geological boundaries by experienced geologists, in particular where data are sparse or of poor quality. The geologist draws such boundaries based on his/her experience and observation of the geology in this area. The final built 3-D geological model of the East Coast area of Singapore is shown in Figure 2.29.

Chapter 3

Planning for Intrusive Ground Investigation

3.1 Adopting the Scope

As stated by Head (1986), site investigation is part of the design process and it is key to an economic and safe design. Head (1986) also stated that successful investigations can only result from thorough planning and design of scoping, for which allocation of realistic funds to enable sufficient planning and implementation of ground investigation are required. The general scope of ground investigation is understood to be as follows:

- profiling of underlying soils and rocks, sampling and testing of soils and rocks
- obtaining groundwater information to assist in obtaining geotechnical parameters for the design

It is also stated in BS 5930 (1999) that the purpose of investigation is to obtain reliable information to produce an economic and safe design, to assess any hazards (physical or chemical) associated with the **ground**, and to meet tender and construction requirements.

It is also suggested that investigation should cover all **ground** in which significant, temporary, or permanent changes may occur as a results of work:

These changes include the following:

- Changes in stress and associated strain
- Changes in moisture and associated volume changes
- Changes in groundwater level and flow pattern
- Changes in property of ground such as strength and compressibility

Before starting the ground investigation, a suitable scope for the investigation should be planned based on the type of infrastructure which is proposed to be developed. The types of structures could be various types of buildings such as residential, commercial, institutional to roads, railways, bridges, ports, airports, municipal services, power stations and associated transmission lines, telecom towers, dams, hydropower dams and stations, land reclamation, engineered landfills, mines and mine infrastructures, and many more.

Adopting of the scope should depend upon the scale of the project, starting from preliminary to detailed ground investigation. Most large-scale projects commence from a preliminary large-scale investigation and then zoom down to a detailed investigation based on the information gathered from the broader ground investigation.

A large-scale investigation generally consists of a preliminary investigation which could cover a large area of interest through aerial photography, remote sensing, digital photography survey, and large-scale airborne to surface geophysical surveys.

The scope should detail the type of investigation at each stage of the phase such as the type of large-scale survey to the type of intrusive ground investigation depending upon the type of infrastructure development. In this scoping study, intrusive ground investigation will be discussed in detail. There are three general types of infrastructure developments such as vertical, linear, and area-wide infrastructures. In general, the ground investigation for vertical infrastructure such as multi-story buildings and telecom towers will emphasize profiling of vertical ground variation in depth assigning a few numbers of ground investigation locations. The linear infrastructure projects such as roads, railways, utility lines, and transmission lines will emphasize the profiling and linear variation of the ground along the proposed route. Area-wide infrastructure such as reservoirs, land reclamations, and landfills emphasizes both profiling and areal variations of the ground across the large area.

When planning the scope, different types of *in-situ* testing, sampling, and laboratory testing should be planned to be able to gather the required geotechnical information depending upon the potential issue such as bearing capacity, stability, settlement, and seepage. The scope of the ground investigation should include a minimum of the following items. Details of these will be discussed in the following sections.

3.2 Adopting Strategy of Ground Investigation

The method of investigation should be selected based on the scale of the project. Large-scale projects involving a large area of investigation are usually carried out under several phases, while small-scale projects within a small area, extent, and/or a simple straightforward project could directly go for a single-phase intrusive ground investigation.

Examples of large-scale investigation are generally carried out applying airborne or surface geophysical mapping as the Phase 1 of the investigation. Intrusive ground investigations are carried out based on the information obtained from the large-scale investigation which could potentially provide a general variation of sub-surface profile and characteristics.

For complex large-scale projects, even the intrusive ground investigation may be carried out over two to three phases. In most of the large-scale projects, investigation of locations with wider spacings is initially carried out as Phase 1 of the intrusive ground investigation, followed by filling up the gap with some additional suitable intermediate locations based on the variation revealed from large-scale Phase 1 ground investigation as Phase 2 of the investigation. For complex projects, Phase 3 investigation applying specialized *in-situ* testing may be carried out to determine the specialized geotechnical parameters.

BS 5930 (1999) generally differentiated three different types of Ground Investigation Categories into Category 1 to 3. The definitions of the categories are provided in Table 3.1 depending upon the type of infrastructure and complexity of the ground condition. The features of categories are described in detail in Tables 3.2–3.4.

Table 3.1 System of site investigation categorization (after Head 1986).

Category	Structure	Ground Conditions
1	Small Simple Straightforward	Uniform, Adequate characteristics
2	Conventional	Varied
3	Large Unusual	Complex Problematic Poor characteristics

Table 3.2 Features of Category 1 investigation (after Head 1986).

Structure Type	Low rise, Light Loaded	Includes housing, industrial buildings, offices, and shops.
Foundation Design	Routine	Based on empiricism and engineering experience, Simple formulae appropriate.
Effect on Environment	No Risk	Will not affect adjacent structures or utilities.
Ground Conditions	Known, Original Materials	Will have adequate strength and low compressibility.

Table 3.3   Features of Category 2 investigation (after Head 1986).

Structure Type	Conventional, Standard foundations	No abnormal loads. Includes buildings, retaining walls, embankments, roads, and bridges.
Foundation Design	Formal Design, Quantitative data required	The foundations should be engineered and be based on experience and judgment.
Effect on Environment	Avoid or accommodate risks	Careful study required.
Ground Conditions	Partly Known, Varied	Field and laboratory tests required. May be specialized investigation aspects.

Table 3.4   Features of Category 3 investigation (after Head 1986).

Structure Type	Unusual Complex Foundations	Abnormal loads and abnormal risk can be identified. Includes high-rise structures, reservoirs, dams, deep excavations, underground works, large-scale civil works, and offshore constructions.
Foundation Design	Specialized, Multiple Aspects, Interactive soil structure design	Could be multidisciplinary.
Effect on Environment	High Risk	Affects life or adjacent property.
Ground Conditions	Unknown, Difficult/Problematic Specialized	Specialized activities would probably be included.

3.3 Selection of the Locations of Holes for Investigation

Selection of numbers and locations of boreholes was briefly described in BS 5930 1999 and Das & Sivakugan (2017). The selected numbers and locations of holes for investigation should reveal the general profile of the ground and variations across the area of the site. When borehole locations are considered, information collected from such boreholes is sufficient to solve the issues of the foundation (either shallow or deep foundation), the retaining structure which may be required for single or multi-level basements, groundwater issues, potential, temporary, and permanent dewatering, seepage problems, any settlement issues, frost penetration and heaving, and seismic activities and earthquakes. Selection of investigation locations for linear projects like highways, railways, transmission lines, and service pipelines may require consideration on cut and fill, embankment stability and ground improvement, and water crossing such as bridges and culvert foundations. The selection of investigation locations for area-wide projects like land reclamations, reservoirs, and landfills may require consideration on overall total and differential settlements, cut and fill exercises, stability of edges, and ground improvement.

The selection of the spacing of boreholes is determined based on the following criteria as stated in BS 5930 (1999):

- Depending on the geometry of structure
- Similarity or complexity of the underlying geology

Brief recommendations on typical spacing of investigation locations are discussed for various infrastructure developments in the following sections as an example. Site-specific considerations and arrangements may require planning for actual ground investigations.

3.3.1 *Residential, commercial, and institutional buildings*

For the simple residential low-rise project, five boreholes are generally sufficient, one at each corner and the remaining one at the center of the footprint as shown in Figure 3.1. Some standards such as BS 5930 (1999) recommended a minimum of three locations. If the infrastructure development involves small multiple units, one investigation location per unit as shown in Figure 3.2 will generally suffice (BS 5930 1999). However, if the footprint of each unit in a multi-unit development is large enough, each unit should consider the approach recommended for a simple residential unit.

The complex multi-story with a bigger footprint area, involving a few levels of basement, may require several boreholes with a suitable grid spacing within the footprint area and investigation locations with a suitable spacing along the edge of

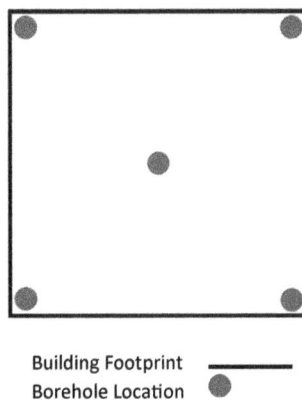

Building Footprint   ——————
Borehole Location   ●

Figure 3.1   Typical arrangement of boreholes and their locations for residential development.

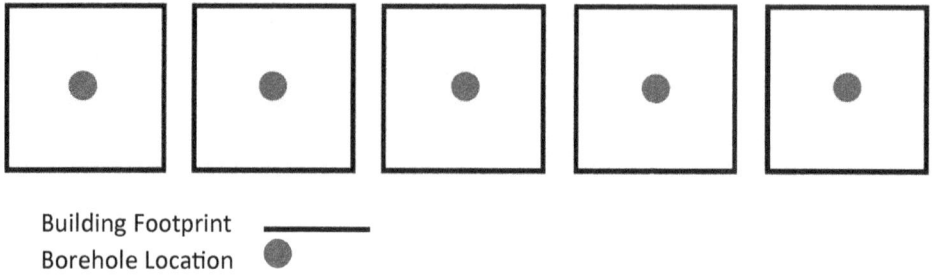

Building Footprint ⎯⎯⎯⎯
Borehole Location ●

Figure 3.2   Typical arrangement of boreholes and their locations for multi-unit residential development.

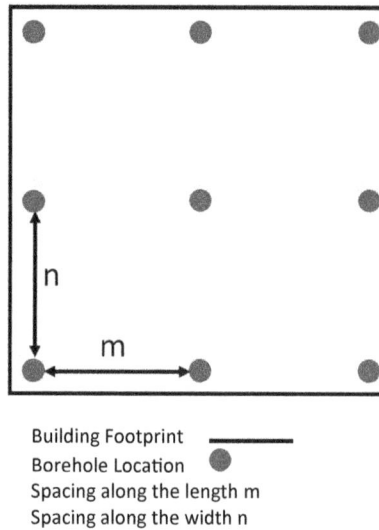

Building Footprint ⎯⎯⎯⎯
Borehole Location ●
Spacing along the length m
Spacing along the width n

Figure 3.3   Typical arrangement of boreholes and their locations for a mid-size, multi-story development with basement.

Note: Spacings n and m can be the same ratio as length and width of footprint.

the footprint to obtain necessary information for the foundation, retaining structure, and necessary hydrogeological information for the dewatering process which may be required. If a deep foundation is required due to greater loading or poor ground conditions, it will be worthwhile considering some borehole locations to be at the potential deep foundation locations, which will also likely be in a grid pattern. The suitable spacing described above may be determined based on the dimension of the footprint and sub-surface information obtained from the large-scale investigation carried out as a preliminary investigation. A typical example of the recommended investigation location layout is shown in Figures 3.3 and 3.4, in

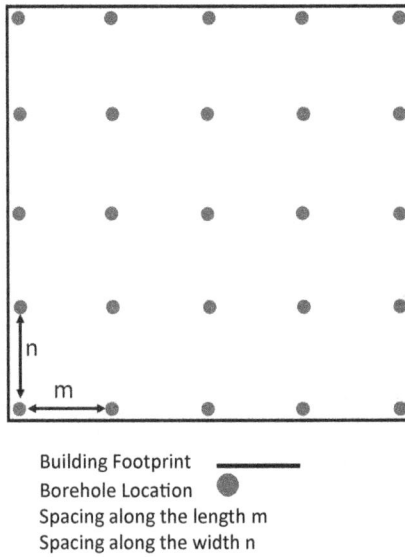

Figure 3.4   Typical arrangement of boreholes and their locations for large size, multi-story development with basement.

Note: Spacings *n* and *m* could be between 1 and 3 times the intended column spacing depending upon the size of the footprint.

which spacing *m* and *n* could be the same for a square footprint. For rectangular spacing, the ratio of *m* and *n* could be the same ratio as the length and width of the footprint. Dimensions of *m* and *n* should be between 0.1 and 0.5 of the length and width with a smaller multiplier being used for a larger footprint area. A similar approach as described for larger residential developments may be required for commercial buildings with a large footprint in which planning of investigation of boreholes with a suitable grid pattern may also be required as shown in Figure 3.4. In this, spacings *n* and *m* could be between 1 and 3 times the intended column spacing depending upon the size of the footprint. However, these recommendations are simple rule-of-thumb recommendations and in practice it could be varied depending upon the complexity of the ground variations, loading conditions, and the geometry of the footprint.

3.3.2 *Highways, railways, transmission lines, and pipelines*

The ground investigation for most linear projects like highways, railways, transmission lines, and pipelines may require investigation locations with regular spacing to profile and obtain geotechnical parameters along the route. For transmission lines, investigation locations are better suited to be placed at the proposed or potential transmission tower locations to collect the geotechnical parameters for the foundation design, which would require both bearing capacity and resistance to lateral and uplift forces.

Projects like highways and railways may require cut and fill exercises, water, valley, and mountain crossings that may require special attention in order to gather geotechnical information for embankments, bridge foundations, and culvert foundations. Borehole locations for highways and railways are generally staggered to cover both edges and centerline locations. Most highway and road agencies around the world have recommended borehole spacings for their regions and areas based on their local experiences. Typical arrangements of borehole locations are shown in Figures 3.5 and 3.6 as an example. Spacing should be based on the total length of the project and expected variations of the ground and groundwater conditions. Most embankments may require investigation locations at the centerline of the embankment for settlement purposes and at the crest of the embankment and the toe of the slope for slope stability purposes as shown in Figure 3.9. For bridge crossing investigations, locations at potential intermediate foundations and abutments may be required. Figure 3.7 shows typical example of borehole layout along utility line investigation.

Borehole spacings, depth of investigation for existing road alignment, new alignment, road widening, and pipeline and culvert foundation investigations are described in detail in the Provincial Pavement Engineering Investigation guidelines version 1.1, 2013, published by the Ontario Ministry of Transportation, Canada.

Figure 3.5   Typical arrangement of boreholes and their locations for a designated highway segment.

Figure 3.6   Typical arrangement of boreholes and their locations for a designated railway segment.

Figure 3.7   Typical arrangement of boreholes and their locations for a designated utility line segment.

3.3.3 *Dams, landfills, and wastewater lagoons*

Planning for investigation locations for dams, landfills, wastewater lagoons, etc., may require both linear and grid pattern approaches. While investigation locations

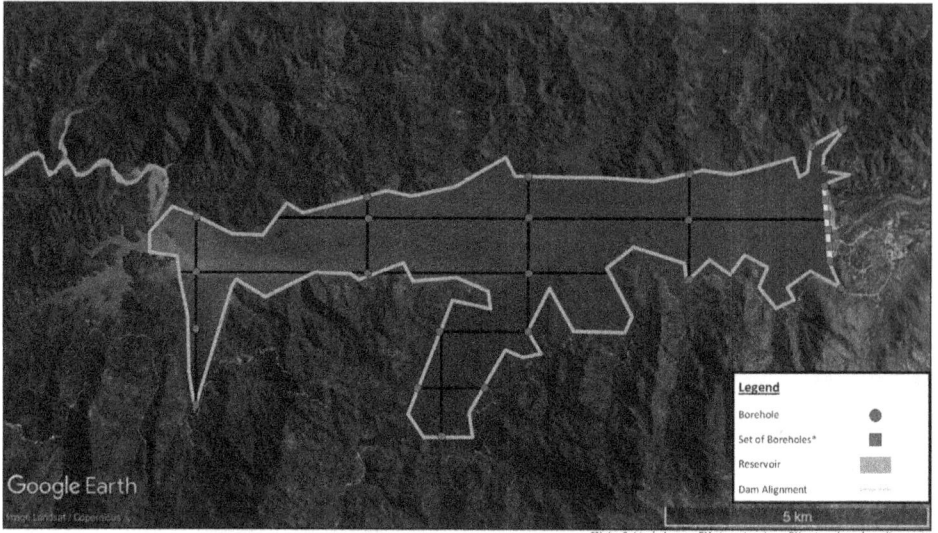

Figure 3.8   Typical arrangement of boreholes and their locations with respect to a reservoir and dam.

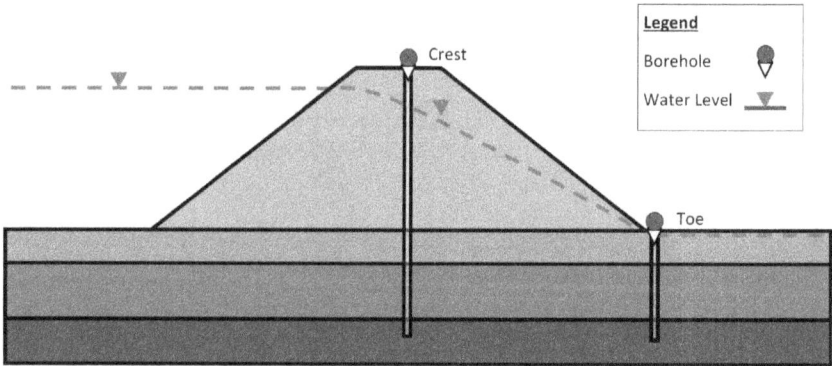

Figure 3.9   Typical arrangement of borehole locations with respect to an embankment dam; recommended locations at the crest and toe of the dam.

at the crest and toe of the slopes with a suitable spacing along the dam, landfill, and wastewater lagoon boundaries for slope stability, some boreholes in a suitable grid pattern within the storage area are needed to determine potential seepage, settlement, etc., as shown in Figures 3.8 and 3.9. Spacings are again based on the size of the project and expected variations of the existing and finished design topography and expected ground variations.

3.3.4 *Land reclamation*

An approach to determine investigation locations for land reclamation may be similar to landfills. Boreholes along edges may be required for slope stability or retaining structure designs, whereas boreholes in a grid pattern may be required for settlement analyses and ground improvement designs as shown in Figures 3.8 and 3.9.

3.3.5 *Pavement for parking areas and driveway*

A few shallow boreholes may be required along driveways with a linear spacing approach, whereas a grid pattern approach may be required for parking lot areas.

In all the above approaches, the dimension of spacing either in the linear approach or the grid pattern may be depending upon ground variations which might have been revealed from preliminary large-scale investigations and/or existing ground information and available budgets.

3.4 Determination of Depth of Investigation

The determination of the depth of investigation for infrastructure development was generally recommended in BS 5930 (1999) as follows:

- The depth to which new structures will affect the ground and groundwater or be affected by them
- Taken below all deposits that may be unsuitable for foundation purposes (e.g., made ground, soft compressible soils).
- If shallow rocks are found, a penetration of at least 3 m in more than one borehole may be required to establish whether bedrock or boulder has been encountered.

The borehole termination depth within overburden is normally determined using Standard Penetration Test blow counts. Termination criteria are generally specified as three continuous blow counts of 50 or 100 per 300 mm penetration whichever is applicable. Where bedrock confirmation is required, one or two core runs of 3 m length in total are normally specified to make sure the bedrock is not to be mistaken with a rock boulder (BS 5930 1999).

In general, the estimated depth could be determined based on the historical borehole information and geological maps. Alternatively, a suitable investigation depth, such as estimated influence depth, can be determined or estimated from stress influence calculation using the type and dimensions of the foundation or infrastructure. For a shallow foundation, influence depth due to foundation could be estimated applying stress influence formulae such as the 2:1 method (Eq. (3.1)) and the Boussinesq 1885 method (Eq. (3.2)):

$$\Delta\sigma_z = \sigma_0 * BL/(B+z)(L+z) \tag{3.1}$$

where Δs_z is the stress increase at depth z,
σ_0 is the initial stress,
B, L, and z are width, length, and depth, respectively.

$$\sigma_z = q_0 * I \tag{3.2}$$

where σ_z is the stress at depth z,
q_0 is surface or contact stress,
I is an influence value which depends on m and n,
m and n values can be found in the U.S. Navy 1971.

For the deep foundation, determination of a competent foundation, such as fresh rock formation which can be safely founded as the end-bearing formation, may be deemed necessary as a suitable investigation depth.

The approximate minimum depth for a building foundation was recommended by the American Society of Civil Engineers (1972) as the depth at which the stress increment is equal to 1/10 of the estimated net contact stress of the foundation (q_0). Sowers and Sowers (1970) also proposed the following equations to determine depths of investigation for hospitals and office buildings:

$$D_b = 3 \times S^{0.7} \text{ (for light steel or narrow concrete building)} \tag{3.3}$$
$$D_b = 5 \times S^{0.7} \text{ (for heavy steel or wide concrete building)} \tag{3.4}$$

where D_b is the depth of boring and S is the number of stories.

In addition, some of the buildings may have one or few levels of basement below the existing or proposed finished grade level. As such, the borehole extending well below the excavation level for the basement will be required. In addition, if there is a potential for an existing groundwater aquifer which has significant piezometric pressure, an investigation of the hydraulic characteristics of such an aquifer may be required in order to obtain the parameters for a hydraulic uplift design. As such, sufficient penetration into the groundwater confined aquifer is required.

A large-area fill such as land reclamation and high embankment may have stress influence reaching to a substantial depth or possibly infinity. As such, the investigation depth extending to non-compressible formations such as the bedrock or very hard or very dense competent formations measuring Standard Penetration Blow counts of 100 per 300 mm for three consecutive penetrations may be required. Existing borehole information and/or existing geological information may indicate such depths.

For land reclamation, embankments, dams, and landfill constructions, penetration depths beyond the potential slip surface of failure may be required for boreholes along the edges. Such slip surfaces are generally extended to the competent formation.

For shore protection structures, earth retaining structures such as sheet pile walls may also require penetration depths well into the competent formation.

For municipal service lines such as water and sewer, it may have to be buried below the influence depth and/or below the frost penetration depth in the cold region to prevent deformation caused by the frost heave.

3.5 Selection of Method of Drilling

There are two main methods of drilling as follows:

(1) Overburden drilling
(2) Bedrock drilling

The method of overburden drilling used usually varies from one country to another. While most of the countries in Asia use a rotary drilling method that uses drilling fluid, the United Kingdom uses percussion methods and the North American region uses the hollow stem flight augering method. Even while using the same method of overburden drilling, engineers and drillers need to plan for a feasible method of penetration as well as supplementary drilling fluid, which may be required depending upon information such as the type of soils and likely potential groundwater pressure. For example, penetration through formation with groundwater pressure may require the application of water balancing, the act of applying heavy drilling fluid, whereas penetration through large gravel and boulder formation may require the application of the water jetting method. A drilling method similar to "down the hole hammer" may be required from time to time to penetrate through large boulder formations or highly fractured rock formations. Some suggested methods of drilling are shown in Table 3.5.

Bedrock drilling is generally carried out applying coring methods to obtain core samples. For most geotechnical investigations, standard coring techniques using single or double core barrels are sufficient, while wireline coring with triple tube core barrels may be suitable where rock characterization is required for extended length.

Based on the type of formation likely to be encountered, suitable drilling equipment is necessary to be mobilized as well as selection of a suitable method of drilling. Details of the drilling method used in geotechnical investigations are described in Chapter 5.

Table 3.5   Recommended methods of drilling for various types of formation.

Type of Formation	Recommended Method of Drilling
Overburden	• Rotary Direct Circulation methods with mud or water flush • Percussive • Auguring
Bedrock	• Rotary Coring • Down the hole hammer
Aquifer with high groundwater pressure	• Rotary Direct Circulation mud flush with heavy drilling fluid or auguring with heavy drilling fluid balancing
Aquifer with high drilling fluid loss	• Rotary Direct Circulation with foam or equivalent
Formation containing large amount of gravel with groundwater pressure	• Water jetting
Formation with boulder or fracture rocks	• Down the hole hammer
Municipal Waste and contaminated land	• Rotary reverse circulation air flush

In addition to the method of drilling, a selection of a suitable type of mounting base for the drill rig will be required depending upon the access conditions. It is necessary to determine the type of mounting system depending upon the type of topography and regime. Most remote access areas may require a drill rig mounted on a skip, tractor, or crawler, whereas truck- or trailer-mounted drill rigs would be suitable to use for ground investigations along an existing road/highway. A portable drilling rig is suitable for areas where access is difficult. These should be highlighted to the drilling contractor during the procurement process.

3.6 Selection of Method of Sampling

During the drilling, soil, rock, and groundwater sampling is usually carried out. Soil sampling is carried out using several types of sampler based on levels of quality required. Sludge samples and disturbed samples are generally required for classification purposes, whereas undisturbed samples are required for strength and consolidation tests to obtain the geotechnical parameters required for the geotechnical design. Depending upon the type and complexity of the project, it is determined whether disturbed or undisturbed samples are to be collected.

Table 3.6   Recommended samplers for various types of soil.

Type of Soil	Type of Sample	Recommended Type of Sampler
Granular soil	Disturbed	SPT split spoon sampler
Non-cohesive silt	Disturbed	SPT split spoon sampler
Soft to Stiff Clay	Undisturbed	Thin Wall Shelby Tube
Soft to very soft clay	Undisturbed	Piston sampler
Hard Clay	Relatively undisturbed	Thick Wall sampler or drive sampler
Organic soil	Relatively undisturbed	Piston sampler
Soft to very soft soil for specialized testing	Undisturbed	Large diameter Laval or Sherbrooke samplers
Soft to stiff clay and sand	Disturbed	Window and windowless samplers

Table 3.7   Sampling interval with Standard Penetration Sampler (recommended in Ministry of Transportation Ontario tender document 2013).

Depth	Sampling Interval
Within the critical zone	Every 0.75 m
Up to 20 m	Every 1.5 m
20 m and above	Every 3 m

Most of the disturbed samples are collected through standard penetration tests or from the sludge or auger returns. Most complex projects requiring complex analyses will require undisturbed sampling. Details of sampling methods are described in Chapter 7.

Rock sampling is generally carried out using a core barrel. Groundwater samples are usually collected from the temporary or permanent groundwater standpipes for further testing. The type of sampling usually used for different types of soils is shown in Table 3.6.

The number of samples to be collected is determined based on the number of formations likely to be encountered and the complexity of the formation characteristics. Most complex projects collect samples for every meter. Some countries, such as Canada, recommend specified sampling intervals for Standard Penetration depths as shown in Table 3.7.

3.7 Selection of Method of *In-situ* Testing

There are several types of *in-situ* testing to determine *in-situ* strengths, consolidation, hydraulic conductivity, and stiffness measurements. Some *in-situ* tests measure direct measurements of geotechnical parameters, while other tests allow geotechnical parameters to be interpreted from the measured parameters applying empirical or theoretical correlations.

Two of the most common types of *in-situ* tests, which are usually carried out together with geotechnical drilling, are Standard Penetration Test and Field Vane Shear Test. While the Field Vane Shear test measures the undrained shear strength directly, SPT measures the blow count over 300 mm penetration called SPT blow count abbreviated as N. From which either undrained shear strength for cohesive soil or relative density and drained friction angle for granular soil could be interpreted.

Depending upon the complexity of the infrastructure proposed for development and the sub-surface conditions, specialized *in-situ* testing, such as Cone Penetrometer, Pressuremeter, and Dilatometer testing, may be required to collect and interpret advanced geotechnical parameters. The types of commonly used *in-situ* testing methods available for geotechnical testing to obtain basic geotechnical parameters can be found in Table 3.8. Details of various types of *in-situ* tests and their applications are described in detail in Chapter 6.

Table 3.8   Types of commonly used *in-situ* testing in Geotechnical Investigation.

Type of *in-situ* Test	Suitable Type of Soil	Obtainable Geotechnical Parameters	Type of Measurement
SPT	Clay, Silt, Sand	• Undrained Shear Strength • Relative Density • Drained Friction Angle for granular soil	Indirect
FVT	Clay and any cohesive soil	• Undrained Shear Strength	Direct
Dynamic Cone	Granular soil, gravel	• Relative density • Drained Friction angle	Indirect
CPT	Clay silt, sand	• Undrained Shear Strength • Drain Friction angle for granular soil	Indirect
Pressuremeter	Clay, silt, sand	• Classification of soils • Undrained Shear Strength • Drain Friction angle for granular soil • Modulus of Elasticity	Indirect
Dilatometer	Clay, silt, sand	• Classification of soils • Undrained Shear Strength • Drain Friction angle for granular soil • Dilatometer Modulus	Indirect
Falling or Constant Head Tests	Clay, silt, sand	• Hydraulic conductivity	Indirect
BAT Permeameter	Clay, silt, sand	• Hydraulic conductivity	Direct

3.8 Services and Archaeological Locations

Before field investigations involving excavations, penetration, and/or drilling are carried out, it is imperative to make sure that valuable sub-surface features and buried utility service lines are not affected by such activities. In order to make sure of this, investigation locations are positioned away from buried utility service lines, and the locating of public and privately owned buried utility service lines, as well as possible buried archaeological features, is carried out. Many countries make utility service location maps available to the public to be able to locate buried utility service lines such as power, gas, municipal sewer, and telecommunication lines when required.

In Ontario, Canada, an organization called Ontario ONE CALL helps in requesting the required locating services for various underground utilities service owners including but not limited to water service lines, gas lines, and Bell and Rogers Canada telecommunication cable lines. Drawings and documents depicting the locations of underground utilities are forwarded to the client once the service providers have located the services.

When requesting locate services, a brief project description is needed to understand the scope of work. The information must include the location of work, the extent of work, and the reason for work.

Next, the utilities owners such as those mentioned above will process the request and commence locating their utilities with respect to the proposed project location. This study is conducted through available resources and possibly physical scanning of the actual locations as well.

Once studies are complete, the gathered documents will be forwarded to the requestee so that they can ensure that their proposed work avoids any located utilities. If needed, the proposed work will be reevaluated to avoid such encounters with located utilities.

Some private utility services which are unlikely to have recorded or registered, utility service locate using magnetic detection or site scan sonar using a private company could be carried out. Most of the municipal services buried nowadays have a metal strip attached to be able to locate using a magnetic detector.

Many archaeological departments maintain known archaeological featured records.

3.9 Planning for Necessary Monitoring and Measurements

During the ground investigation stage, it may be necessary to gather measurements with regard to encountered groundwater and gas for use in design and remediation plans. In order to obtain measurements, installation of a set of groundwater and gas monitoring wells is needed so as to monitor and sample encountered elements. These requirements need to be planned together with the ground investigation so that some of the ground investigation boreholes could be used to install such monitoring wells.

Details of geotechnical instrumentation during ground investigation are described in Chapter 8.

Chapter 4

Method of Intrusive Ground Investigation and Procedures

There are various methods of intrusive ground investigation to reveal the subsurface ground and groundwater conditions. Some of them are suitable for shallow investigations and others are suitable for relatively deep investigations. Depending on the type of advancement, some are suitable for overburden soils, some for very dense to very hard overburden soils to bedrocks, and yet others for relatively softer or looser soils. Some may be suitable for cohesive soils and others may be suitable for granular soils. Formation bearing groundwater and groundwater pressure may call for special techniques of controlling borehole stability and collapse. Details of various advancement techniques are described in the following sections.

4.1 Test Pits

Test pitting can be carried out with handheld equipment, such as shovels, or machine pits using a backhoe excavator, depending upon the type of the soil expected to be encountered and the purpose of investigation. A hand excavation can be carried out manually for typically up to one meter in depth, whereas machine excavation can be carried out up to 3 to 4 meters without support depending upon the type of soil encountered. Depending upon the type of soil, the necessary gradient of the slope should be provided for stability and safety purposes. The Occupational Health and Safety Procedure for excavation was extensively described in the Excavation section of Construction Regulation in Canada (IHSA 2019) based on the type of soil to be excavated. Generally, granular loose soil requires a sufficiently gentle slope for safety and stability purposes.

Test pits are suitable for quick and shallow investigations and inspections, especially for foundations of shallow footing as well as for services installations such as sub-surface water, sewer, gas, electricity, and telecommunication lines.

Test pitting allows logging of the soil profile, observing seepage and ground-water sampling, as well as collection of disturbed or intact undisturbed samples. Examples of test pit excavation are shown in Figures 4.1 and 4.2.

4.2 Trenches

Trenching is very similar to test pitting and generally uses the excavation machine to excavate the trench. Similar to conventional test pitting, it can excavate up to 4 meters without support. Unlike test pitting, it excavates a long trench along the area of interest. It is also a quick method of investigation for shallow soil profiling which also allows one to collect disturbed and undisturbed samples. Figure 4.3 shows trenching in progress.

4.3 Drilling

There are several methods of drilling to advance the boreholes in order to obtain sub-surface soil and rock profiles and collect soil, rock, and groundwater samples

Figure 4.1   Test Pit excavation using backhoe excavator.

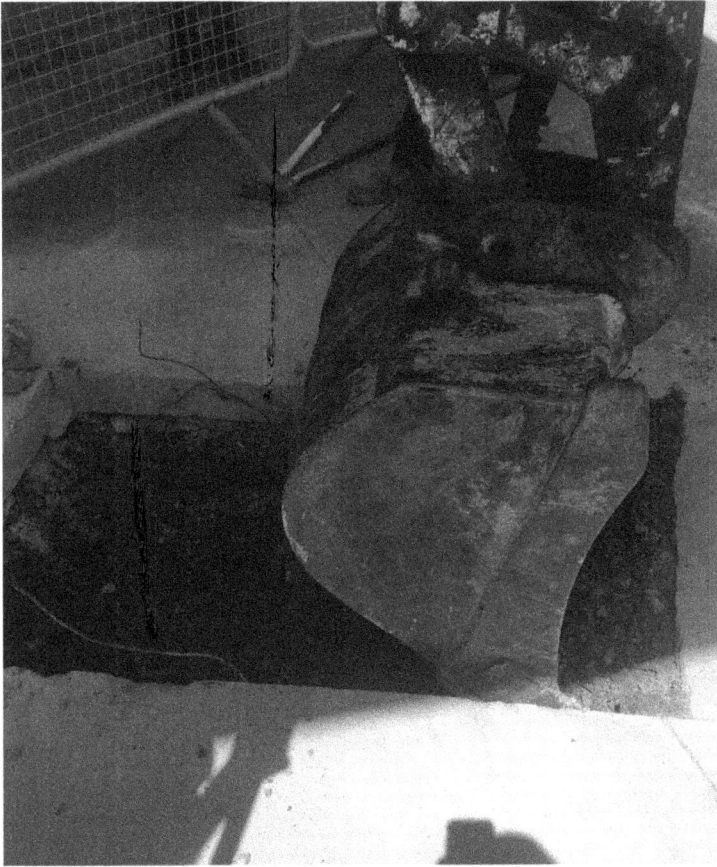

Figure 4.2   Test pitting in progress.

for further laboratory tests for geotechnical characterization. Some of the commonly used methods of advancing boreholes in the geotechnical industry are as follows:

Method of Drilling:

- Wash Boring
- Cable percussive
- Augering
- Hollow stem flight auger
- Rotary open hole drilling
- Coring
- Down the hole hammer

Figure 4.3   Trench excavation using backhoe excavator.

4.3.1 *Wash boring*

The wash boring method advances the boreholes using a powerful water pump and jet in the water through the drill pipe and a drill bit. Through this method, soils are disintegrated and lifted up through the annulus of the borehole to the surface. This method is suitable for advancing boreholes for granular soil such as sand, gravel, and small boulders, especially when high groundwater pressure is encountered. Disturbed sludge samples can be collected from the return water to the surface. Figure 4.4 shows wash boring using a tripod. This method can be used in conjunction with any other drilling method where water or drilling fluid pumping facilities are available. This method is frequently carried out in open holes or cased holes where it is required to maintain stability of the boreholes.

4.3.2 *Cable percussive drilling*

This method is extensively used in the United Kingdom and, to certain extent, in Southeast Asia. This method simply drops a suitable sharp drill bit attached to the drill rod which is pulled and dropped using a steel cable with the help of human or machine power along with gravity. A bailer, circulated water, or drilling fluid is

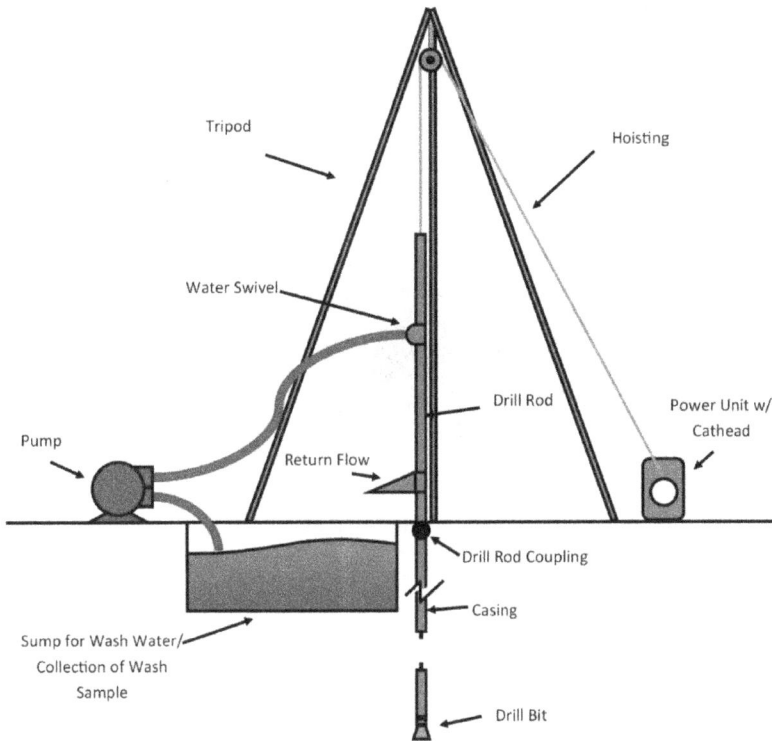

Figure 4.4   Wash boring rig set up.

used to remove the drill cuttings. Disturbed samples can be collected from the bailer or returned drilling fluid. This method is suitable for advancing boreholes through both granular and cohesive soils. Undisturbed samples can be collected at the bottom of boreholes using appropriate samplers. When the formation of granular deposit with groundwater pressure is encountered, one can easily switch to the wash boring method using the same equipment provided a water pump is available. This method is frequently carried out in an open hole or is carried out with the addition of necessary casing behind the drill bit for borehole stability. Figure 4.5 shows drilling using the cable percussive method.

4.3.3 *Augering*

This method advances the borehole by augering the soil. In many cases, this method is carried out under dry conditions in an open hole. Following casing behind the tip of the auger is also possible to be able to maintain the stability of the borehole. Soil samples can be collected from the auger returns. Augering could be carried out using a hand auger, with a hand-operated or machine-operated, or

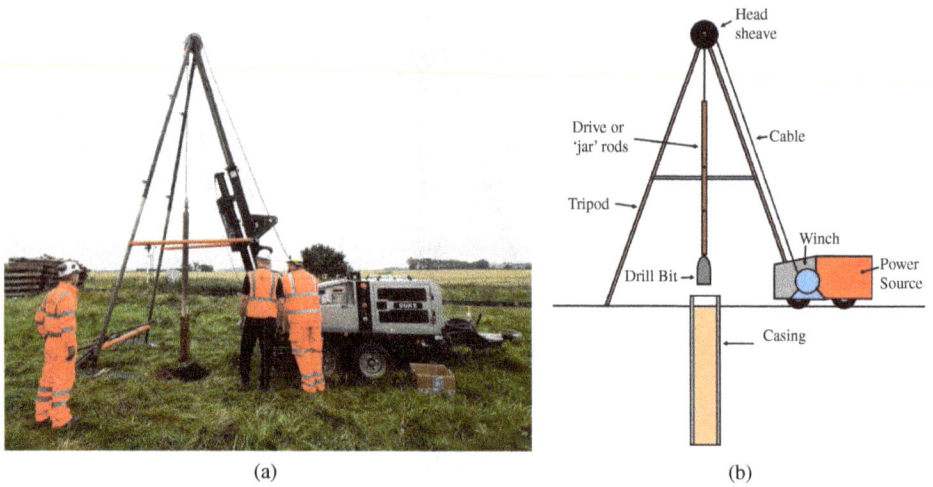

(a)   (b)

Figure 4.5   (a) Cable percussive drill rig (Courtesy of "Dando Drilling Ltd"), (b) Schematic diagram.

Figure 4.6   Augering in progress using solid stem auger.
Source: Dando Drilling Ltd.

drill rig with augering drill rod. There are two types of auger stems known as solid stem auger and hollow stem auger. These augering methods are extensively used in the geotechnical drilling industry of North America. Figure 4.6 shows augering in progress using a solid stem auger.

4.3.4 *Hollow stem flight auger*

The hollow stem flight auger drilling method is extensively used in North America. The benefit of using this method is that the hollow stem acts as a casing to maintain borehole stability. In addition, undisturbed sampling and *in-situ* tests, such as the Standard Penetration Test and Field Vane Test, can be carried out through the hollow stem. Disturbed samples can be obtained from the auger returns as well. In case of encountering groundwater pressure, the application of hydraulic balancing is possible by filling the water head within the hollow stem. Application of the wash boring technique through the hollow stem is also possible. The hollow stem flight auger method can penetrate through overburden soil such as clay, sand, and gravel. Figure 4.7 shows hollow stem augering in progress.

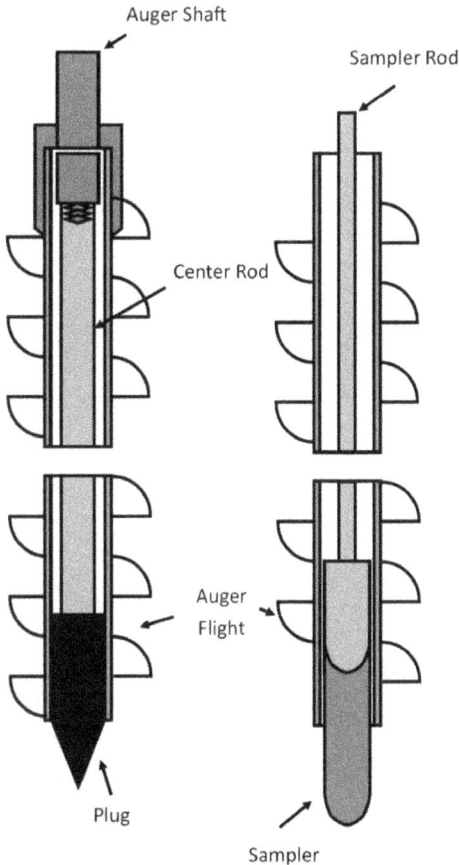

Figure 4.7   Hollow stem flight auger and sampling operation in progress.

4.3.5 *Rotary open hole drilling*

The rotary open hole drilling method is generally used in Southeast Asia. This method allows drilling of a borehole without casing. The stability of the borehole is normally maintained with the help of drilling fluid circulation. This method results in less disturbance to the formation. Soil samples can be obtained from the sludge returns. Alternatively, a collection of disturbed and undisturbed samples can be carried out using a split spoon sampler and various types of undisturbed sampling tubes. This drilling method is considered to be the best method to control groundwater pressure and drilling fluid loss by controlling the characteristic of the drilling fluid with suitable specifications. The standard type of drilling fluid used is generally bentonite slurry. Most rotary drilling is carried out by applying the direct circulation method in which drilling fluid is pumped in through the drill rod and fluid returns via the annulus between the drill rod and the borehole wall. The drilling fluid serves several purposes throughout the drilling process, such as cooling the drill bit and rod, maintaining the borehole wall by means of sealing the pores in the

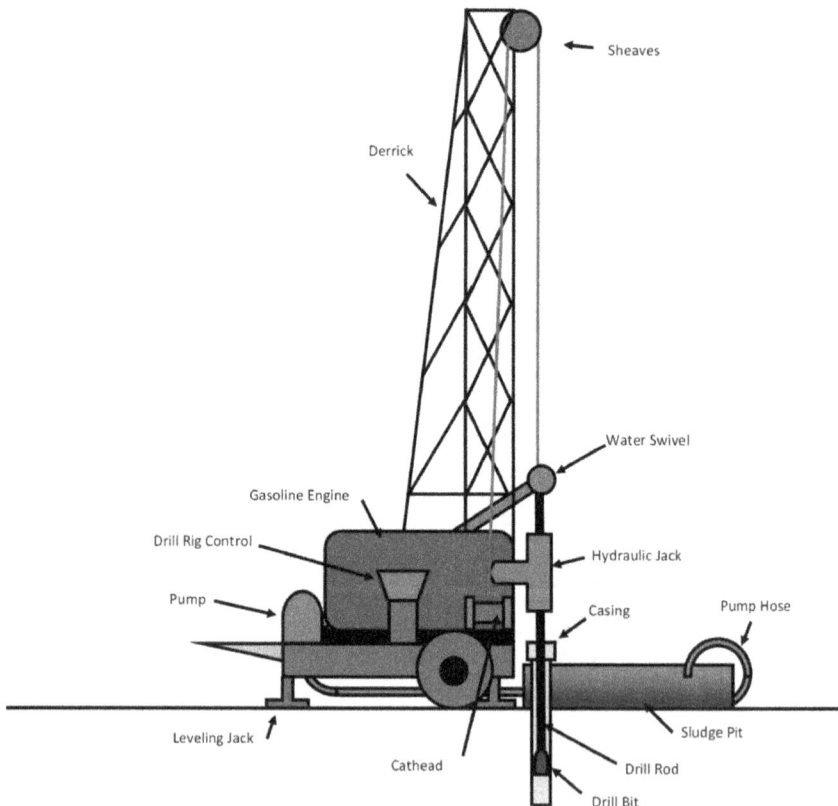

Figure 4.8   Rotary drill commonly used in geotechnical ground investigations.

wall, preventing groundwater pressure, and carrying the drill cutting to the surface. The velocity required to carry the drill cuttings is generally achieved by selecting suitable sizes of drill bits and drill rods as the resulting annulus provides the suitable velocity. Figure 4.8 shows rotary direct circulation drilling.

In some cases, a reverse circulation process is applied in which drilling fluid enters from the annulus and returns through the drill rod. While drilling within a landfill or contaminated land, the reverse circulation method applying an air flush is usually used so as to avoid cross contamination caused by the drilling fluid. In that case, air circulation allows the drill bit and accessories to cool down and also carries the drill cutting by air flow. Figure 4.9 shows the rotary direct circulation and reverse circulation drilling methods.

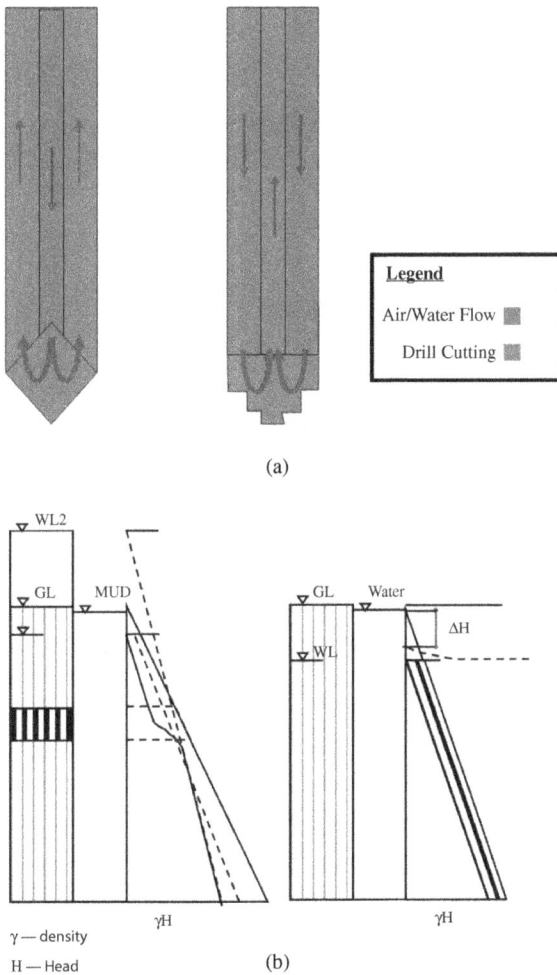

Figure 4.9   Rotary direct circulation: (a) Mechanism of direct and reserve circulation methods and (b) Fluid pressure changes during direct and reverse circulations.

4.3.6 *Coring*

Coring, generally, is used to advance the borehole to penetrate the rock formation. The rotary direct circulation method in combination with the use of a core barrel allows one to cut and retrieve the core samples. Depending upon the type of rock, single-, double-, or triple-tube core barrels are used to achieve the required quality. The specified core barrel is normally attached with a diamond or tungsten carbide bit to cut the core. The length of the core barrel is standardized to 3 meters. In the case where deep penetration within the rock formation is required, wireline coring is commonly used in which the core barrel can be retrieved to the surface within the large-diameter drill rod. Figures 4.10–4.12 show the coring technique, core barrel, and core bits, respectively.

4.3.7 *Down the hole hammer*

Down the hole hammer drilling techniques are generally used to break and penetrate the bedrock formation. Dynamic forces exerted through the tip of the drill

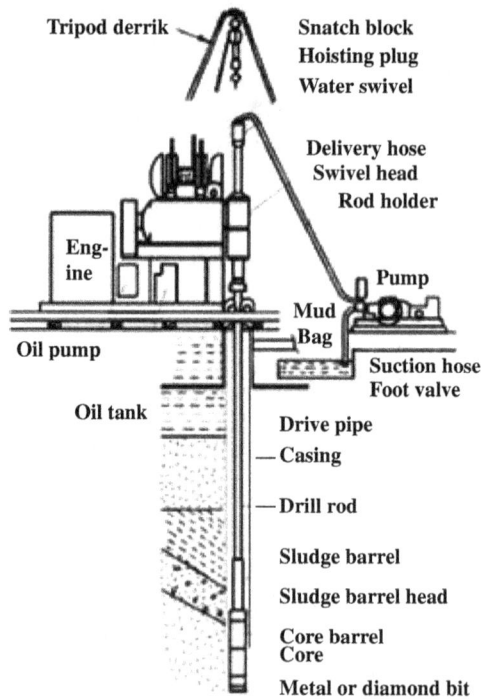

Figure 4.10   Coring technique used in the field.
Source: Bo and Choa (2004).

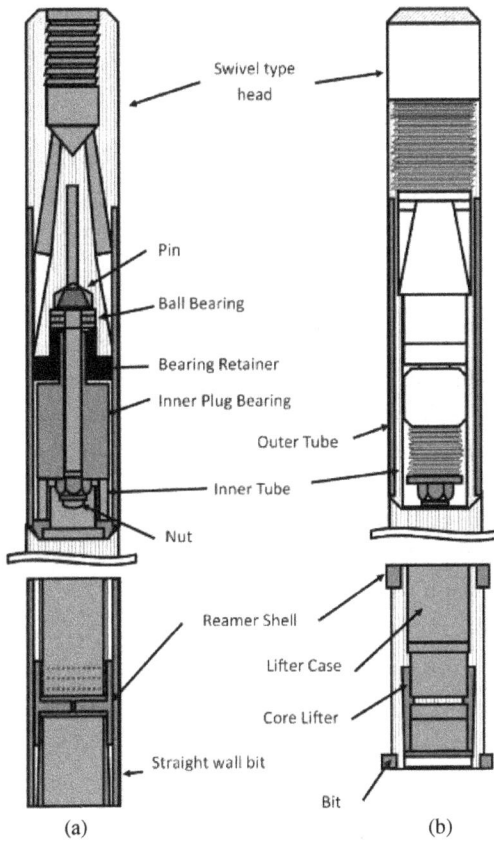

Figure 4.11   Core barrels: (a) double-tube core barrel and (b) barrel and bit.

Figure 4.12   (a) Diamond bits and (b) reaming shells.
Source: Courtesy of TMG Manufacturing.

bit are used to break the rock for which air or water forces are normally applied using an air compressor or pump. In this equipment, a hammer is directly attached to the drill bit. The hammer piston strikes the bit resulting in an efficient transmission of energy with minimal power losses due to introducing power within the hole at a depth. Only disturbed broken rock samples can be obtained in this exercise when the reverse circulation technique is combined with this method. Figures 4.13 and 4.14 show the down the hole hammer rig and attachment.

4.4 Various Types of Drill Rig Mounting

Depending upon the site conditions, topography, and terrain, different types of mounting rigs are required in order to suit the access and mobility of the available sites. The following types of rig mounting on various systems are available in the market:

- Drill rig mounted on a truck
- Drill rig mounted on a trailer
- Drill rig mounted on a tractor

Figure 4.13   Down the hole hammer rig.
Source: Courtesy of Epiroc Canada.

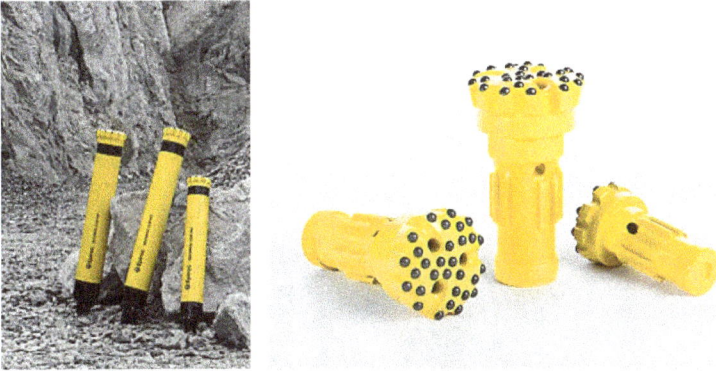

Figure 4.14   Attachment to hammer rig where air forces are used to create dynamic forces.
Source: Courtesy of Epiroc Canada.

- Drill Rig mounted on a crawler
- Drill rig mounted on a skid
- Drill rig with a portable tripod
- Drill rig mounted on a floating pontoon or jack-up pontoon
- Drill rig mounted on a ship

Drill rigs mounted on a truck or trailer are suitable for drilling projects along the highway and projects requiring long-distance mobilization and movement. Drill rigs mounted on a tractor or crawler are usually utilized in remote area projects where the terrain is rough and access is poor. Crawlers consist of two types of tracks such as rubber and steel tracks. Rubber tracks are used where it is imperative to avoid damage to the access road caused by track marks. These types of crawler-mounted drill rigs are required to be transported to a nearby site location through a good access road using a flatbed loader vehicle. Skid-mounted rigs are generally used in remote locations where access and soil conditions at the ground surface are poor. Portable drill rigs with a tripod are also used where access is limited and the area for drill sitting is limited. Figures 4.15–4.19 show drilling rigs mounted on various types of moveable bases.

Drill rigs mounted on pontoons are normally used in investigations in the water and foreshore area (Figure 4.20). Drill rigs mounted on ships are used for most offshore soil investigation projects (Figure 4.21). This type of drill vessel is also usually equipped with an on-board soil laboratory.

4.5 Probing

There are a few methods of probing which can be accompanied with drilling. Most of them apply geophysical technology and many of them have been used in

Figure 4.15   Drill rig mounted on a truck.
Source: Courtesy of CME.

oil and water well industries. These probing techniques, using geophysical logging, are applied in the geotechnical industry from time to time. These are as follows:

- Gamma–Gamma Logging
- Neutron Probe
- Resistivity Probe
- Electric Conductivity Probe

Nuclear gauges consisting of gamma and neutron scattering have been used in the geotechnical industry for earthworks quality control and testing. Soil properties such as unit weight can be interpreted from measured parameters from gamma scattering, whereas moisture content can be interpreted from neutron scattering which measures hydrogen ion content. The methods of measurement are described in detail in ASTM D 2922 and 3017 and BS 1377 (1990). Figure 4.22 shows a nuclear gauge and Figure 4.23 shows the method of

Figure 4.16   Drill rig mounted on a crawler.
Source: Courtesy of CME.

measurement. As the equipment uses radio isotopes, users are frequently required to protect themselves from overexposure to radioactivity from the use of this equipment. The count of exposure is registered through the batch it wears which is tested from time to time to monitor the magnitude of exposure. As the element of radioactivity decays with time, frequent calibration of the equipment is required.

In addition to the nuclear gauge, other types of probes described above can be used to classify the type of soil, groundwater level, and determine their salinity. Three-channel probes consisting of Gamma–Gamma, resistivity, and spontaneous potential measurements are frequently used in formation identification in petroleum and water well industries. While Gamma–Gamma and resistivity profiles can identify and differentiate clay and granular soils, spontaneous potential can identify the salinity of groundwater. Figure 4.24 shows a three-channel geophysical logger and winch (ASTM D6726 and D6727 2007). Details of geophysical logging and the equipment are described in Chapter 7.

Figure 4.17   Drill rig mounted on a skid.
Source: Courtesy of CME.

Figure 4.18   Drill rig mounted on a trailer.
Source: Courtesy of DST Consulting Engineers Inc.

4.6 Conventional *In-situ* Testing

There are a few methods of *in-situ* tests which usually accompany the drilling process as follows:

- Standard Penetration Test (SPT)
- Field Vane Shear Test (FVT)

Another test that is frequently carried out after drilling and installation is

- Hydraulic Conductivity Test

These tests will be described in detail in the following sections.

4.6.1 *Standard Penetration Test (SPT)*

Standard penetration tests are generally carried out within the overburden soil to collect disturbed samples and to obtain SPT blow counts to interpret the

Figure 4.19   Drill rig mounted on a tractor.
Source: Courtesy of DST Consulting Engineers Inc.

Figure 4.20   Ground investigation at foreshore location carried out from a jack-up pontoon.
Source: Bo and Choa (2004).

Figure 4.21   Drilling ship with on-board laboratory built on the ship.
Source: Courtesy of Shutterstock.

Figure 4.22   Nuclear gauge used for density and moisture content measurements.
Source: Courtesy of Matest S.p.A.

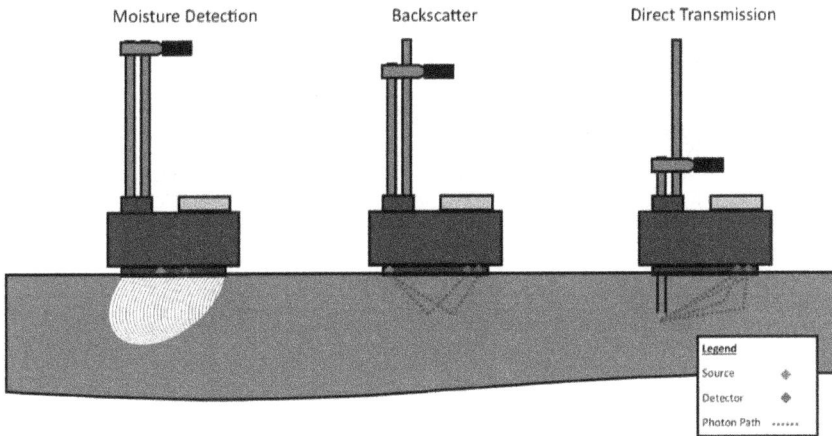

Figure 4.23   Methods of measurement using Nuclear gauge.

consistency of the soil encountered. SPT sampling using a split spoon sampler cannot obtain an undisturbed sample due to application of dynamic forces and the significant thickness of the sampler wall which disturbs the soils. Despite blow count numbers called "*N*" value being used to estimate the relative density and stiffness (consistency) of the soil, it is only useful and valid for highly permeable soils such as granular soils where excess pore pressure generation is limited. It is

Figure 4.24   Geophysical logging equipment and winch.
Source: Courtesy of Mount Sopris Instruments.

Figure 4.25   Hammer used for SPT.
Source: Bo and Choa (2004).

rarely useful for very soft to soft soil as excess pore pressure generation is significant in such soil.

There are some useful empirical correlations between the SPT blow count "*N*"
and basic geotechnical parameters such as Undrained Shear Strength (Terzaghi *et al.*
1996) and Drained Friction angle (Wolff 1995) as shown in Equations (4.1) and (4.2):

$$s_u = 4.4 N_{60} \left(kPa \right) \tag{4.1}$$

$$\Phi = 27.1 + 0.30\,N_{cor} - 0.00054\,N^2{}_{cor} \tag{4.2}$$

where s_u is undrained shear strength,

N_{60} is the SPT blow count corrected for energy at a 60 % level,

Φ is drained friction angle, and

N_{cor} is the corrected SPT blow count.

The Standard Penetration Test is also used to determine the termination depth of drilling. Usually, three (3) consecutive 50 blows per 300 mm penetration are used to specify the termination criteria.

SPT tests are generally carried out by dropping the SPT hammer with a weight of 63.5 kg from a 760 mm height onto the anvil attached to the drill rod. Nowadays, automatic trip hammers are usually used. The blow counts recorded over a penetration of 300 mm after an initial 150 mm of penetration (usually called seating drive) are termed as "*N*" value. SPT tests are usually carried out in accordance with ASTM D1586M (2018) or BS 5930 (1999) and BS 1377-9 (1990). There is also a special standard for carrying out SPT within the hollow stem auger (ASTM D6151M 2015). Figure 4.25 shows an SPT hammer and Figure 4.26 shows SPT testing in progress.

4.6.1.1 *Factors affecting SPT blow counts*

Even with SPT being carried out in the same soil at the same depth and under the same groundwater conditions, the resultant blow count numbers could be affected

Figure 4.26   SPT test in progress.
Source: Bo and Choa (2004).

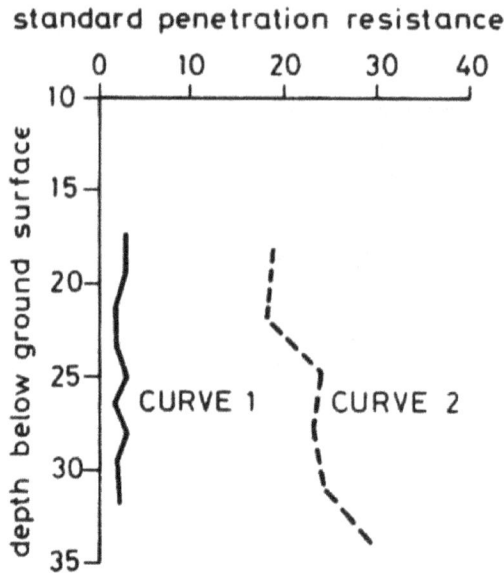

Figure 4.27   Effects of borehole disturbance on SPT results for depths (meters) below ground surface.

Source: Thorburn (1986).

by several factors. SPT tests carried out by two industrial contractors at the same site were compared by Thorburn (1986). As can be seen in Figure 4.27, two significant profiles of SPT were obtained in the sand due to a significant effect of borehole disturbances.

The following are factors affecting the SPT blow count in the borehole:

- Hammer efficiency
- Energy transfer ratio (Length of rod and stiffness)
- Overburden pressure
- Groundwater level in the hole
- Borehole size
- Quality of borehole

In order to compensate for those effects, the followings correction called N value at 60% energy (N_{60}) was recommended:

$$N_{60} = 1.67 E_m C_b C_r N \tag{4.3}$$

E_m = 0.6 for safety hammer, 0.45 for Doughnut hammer
C_b = 1.0 for BH 65-115 mm, 1.05 for BH 150 mm, 1.15 for 200 mm
C_r = 0.75 for up to 4 m drill rod, 0.85 for 4–6 m, 0.95 for 6–10 m, 1 for above 10 m

Overburden correction:

$$N_{cor} = C_N \, N_F \tag{4.4}$$

$$C_N = 9.78 \left(1/\sigma' \right)^{0.5} \tag{4.5}$$

Dilation correction:

$$N' = 15 + 0.5 \left(N_{cor} - 15 \right) \tag{4.6}$$

Figure 4.28 shows how borehole size can affect the SPT blow counts. SPT blow counts were compared with the N value obtained from 150 mm boreholes and boreholes with other diameters. It can be seen that smaller-diameter boreholes produce a larger N value compared to larger-diameter boreholes (Figure 4.28).

4.6.2 *Field Vane Shear Test (FVT)*

Field Vane Tests (FVTs) are carried out with a standard vane with two common types of vane blades such as 55 mm × 110 mm and 65 mm × 130 mm depending upon the strength of the clay encountered. In general, a larger vane is used for softer clay and a smaller vane is used for firmer clay. Vane equipment is inserted either through the open hole or within the casing at the bottom of the hole. In North America, vane equipment can be inserted through a hollow stem auger. The vane blade is usually penetrated to the depth which is four times the width of the vane blade into the soil mass in order to avoid the disturbed zone of soil caused by the drilling process. Certain waiting times, generally 5 minutes after insertion of the blade, and a suitable rotation rate of 0.2 degree per second are used as advised by Chandler (1988). A field vane basically measures the torque required to rotate the blade until failure. Field vane equipment has to be calibrated from time to time. Field vane tests are generally terminated at field vane shear strength of 90–100 kPa. Remolded strength is also usually obtained to remold the soil by rotating the vane blade up to 25 times in full rotation before carrying out the remolded strength tests. Field vane shear tests are usually carried out in accordance with ASTM 2573-08 and/or BS 5930 (1999). Undrained shear strength can be interpreted from measured torque values using the following equations:

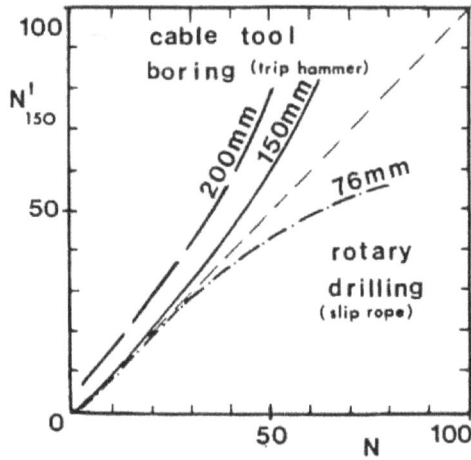

Figure 4.28   Effects of borehole size on SPT. "Comparison of SPTs for alluvial sands".
Source: Thorburn (1986).

$$s_u = M / K \tag{4.7}$$

where
   s_u is undrained shear strength in kPa,
   M is torque to shear the soil in N m, and
   K is constant depending upon the dimensions and shape of the vane.

For the vane which has a height into width ratio of 2,

$$K = 3.66 D^3 \times 10^{-6} \tag{4.8}$$

where
   D is the measured width of the vane.

Figure 4.29 shows field vane blades and the field vane equipment. Figure 4.30 shows typical calibration results of the field vane equipment. The equation 4.7 shows a relationship between the torque and undrained shear strengths of cohesive soils for two types of vanes. Table 4.1 shows the typical presentation of Field Vane Shear Test results and Figure 4.31 shows a graphical presentation of the results.

Bjerrum (1972) suggested that the measured field vane shear strength should be corrected using the correction factors which vary based on the plasticity index as shown in Figure 4.32.

4.6.2.1   *Factor affecting field vane test results*

Like the SPT test, the resulting field vane shear strength could also be varied in the same type of soil at the same depth and under the same groundwater conditions due to several factors which could affect the test results.

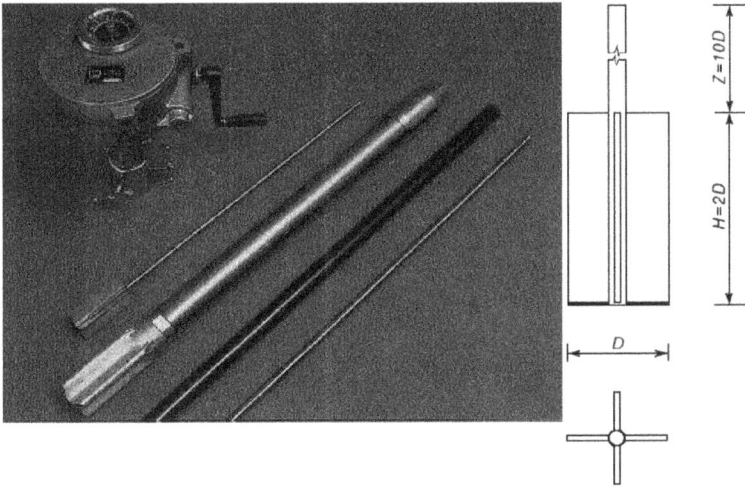

Figure 4.29   Field vane blades and field vane equipment.
Source: Bo and Choa (2004).

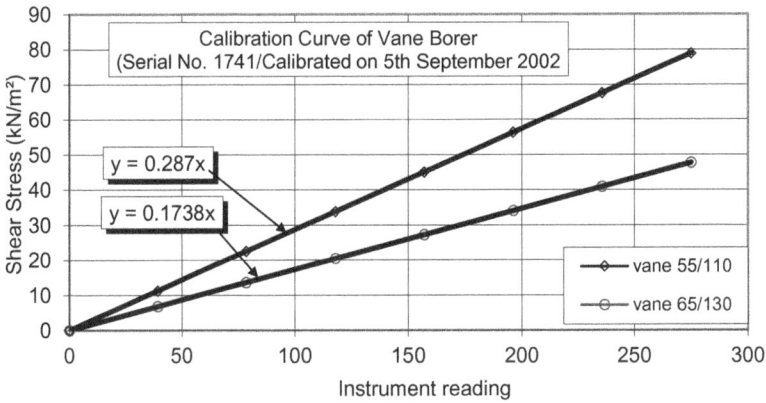

Figure 4.30   Typical calibration results of field vane equipment.
Source: Bo and Choa (2004).

The followings are some factor affecting the field vane test results:

- Waiting time
- Rotation rate
- Dimension and thickness of vane blades

During the insertion of the vane into the cohesive soil, the generation of excessive pore pressure occurs. This pore pressure development and dissipation could

Table 4.1   Typical field vane shear strength results.

Serial No. 1741							05-Sep-2002
		Instrument Reading				**Average**	**Shear Strength (kN/m²)**
Total Load (N)	**Up(1)**	**Down(2)**	**Up(3)**	**Down(4)**		**(1+3)/2**	**Vane 55/110**	**Vane 65/130**
0.00	0.0	0.0	0.0	0.0	0.00	0.00	0.00
39.24	13.0	12.5	13.0	13.0	13.00	11.26	6.82
78.48	25.0	25.0	25.0	25.0	25.00	22.52	13.64
117.72	37.0	36.8	37.0	37.0	37.00	33.79	20.46
156.96	48.5	48.0	48.8	48.8	48.65	45.05	27.28
196.20	60.3	60.7	60.8	60.8	60.55	56.31	34.09
235.44	71.7	72.0	72.0	72.2	71.85	67.57	40.91
274.68	83.0	83.0	83.0	83.0	83.00	78.83	47.73

Source: Bo and Choa (2004).

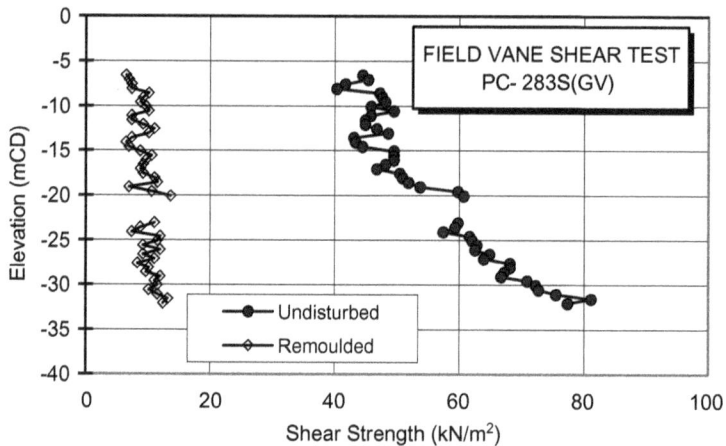

Figure 4.31   Typical field vane shear strength results.
Source: Bo and Choa (2004).

vary the effective stress of the soil hence soil strength measured could be affected. Figure 4.33 shows increasing undrained shear strength values with waiting time between insertion and rotation. The longer the waiting time, the greater the undrained shear strength due to effective stress gain.

Figure 4.34 shows increasing undrained shear strength with an increase in the rate of rotation as the undrained shear strength is not unique and has strain-dependent characteristics.

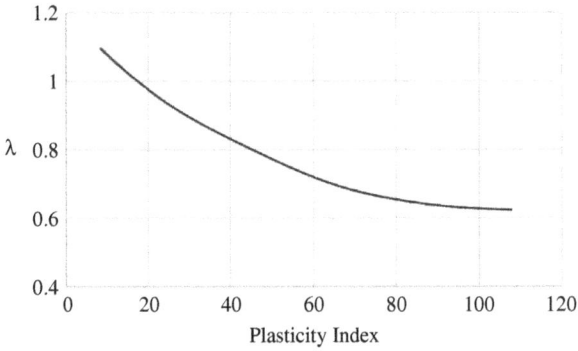

Figure 4.32   Correction factors for field vane strength based on plasticity index.
Source: Bjerrum (1972) with permission from ASCE.

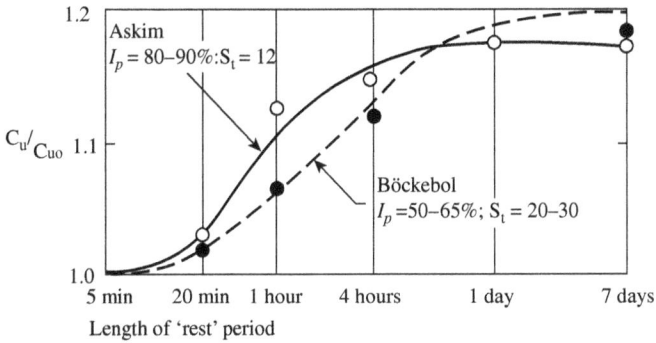

Figure 4.33   Effects of delay between insertion and rotation of field vane on measured undrained strength.
Source: Chandler (1988).

The dimensions of the blade could affect the measured undrained shear strength when using the field vane equipment as the greater the thicknesses, the lower the undrained shear strength. Figure 4.35 shows decreasing undrained shear strength with increasing thickness of the vane blade when measured using the field vane shear equipment.

Therefore, it is recommended to follow standard procedures and use standard dimensions of vane blades to obtain accurate and realistic results from the field vane testing.

Standard testing procedure

- Vane Size: Small vane: 55 × 110 mm (suitable for clays C_u = 30–110 KPa)
                      Large vane: 65 × 130 mm (suitable for clays C_u < 50 KPa)
- Area ratio < 12 %

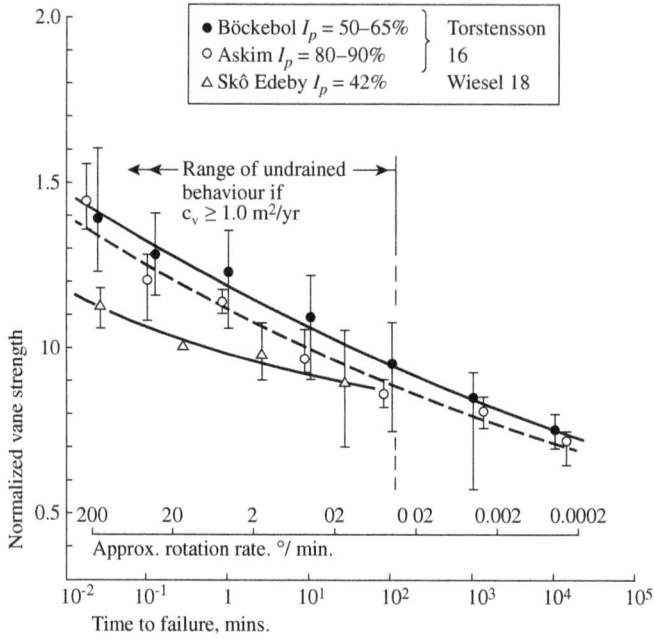

Figure 4.34 Effect of rate of rotation on field vane strength for high I_p clays.
Source: Chandler (1988).

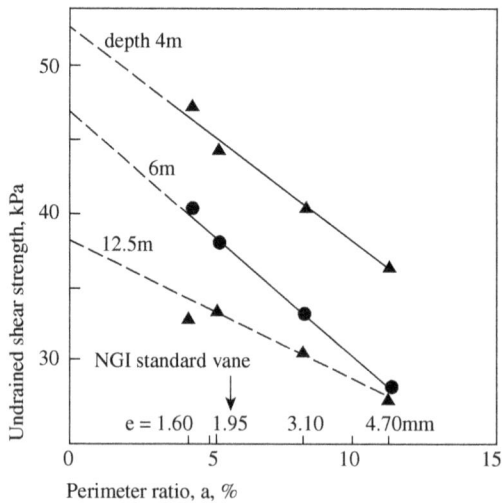

Figure 4.35 Effect of vane blade thickness on measured undrained strength.
Source: Chandler (1988).

- Rod size – OD: 32 mm, ID: 22 mm
- Penetration depth: 4 B here B is width of blade
- Waiting time: 5 minutes
- Rotation rate: 0.2° per sec
- Remolded strength: after 25 full rotations
- Calibration frequency: 6 months.

4.6.3 *Hydraulic conductivity tests*

Hydraulic conductivity tests to determine hydraulic characteristics of soil or rock masses are carried out within an open hole during drilling processes or in a cased hole after construction of monitoring well. Hydraulic conductivity tests are carried out after application of a hydraulic gradient or pressure onto the groundwater or rock fracture, respectively. Hydraulic gradients can be applied as either positive or negative head above or below the existing groundwater head. The following are types of tests carried out to obtain hydraulic characteristic of groundwater aquifer or rock masses:

- Falling head test
- Rising head test
- Constant head test
- Packer test

All of the above tests, except the packer test, are usually carried out to determine the hydraulic conductivity of soil mass. The falling head test applies water head above the static groundwater level to create a hydraulic gradient and measure the dropping of the applied head levels with the selected time intervals. The rising head test applies negative head of water below the static groundwater level to create a hydraulic gradient and measures the rising of groundwater levels under the selected time intervals. The constant head test applies certain constant water head above the static groundwater level and measures the volume of water required to maintain the constant level of head above the static groundwater level within selected time intervals. The time intervals selected are usually closer intervals in the initial period of tests as the rate of dropping or recovery of applied groundwater levels is faster during that period. Time intervals are gradually transitioned to wider time intervals at the later part of the test. The following Tables 4.2 and 4.3 shows field-recorded groundwater levels during the falling and rising head tests and Table 4.4 shows field-recorded data showing a constant head test. Using those data, necessary time factors were interpreted from the graphical analyses as shown in Figures 4.36, 4.37 and 4.38.

Using the time factor interpreted from these test, the following equation can be applied to obtain hydraulic conductivity of soil mass:

Table 4.2   Field hydraulic conductivity test data from a falling head test.

Slug Test Result	Project No.	HG-20201001
Conducted by		Carey P.
Test Location		BH-1
Test Date		6-Jan-20
Testing Depth (m)		4.5–4.7 m
Test Formation		Silty Sand
Initial Ground Water Depth, H (m)		2.67
Starting Water Depth, H_0 (m)		1.54
Time Interval (min)	**Water Level, h (m)**	**$H\text{-}h/(H\text{-}H_0)$**
0	1.54	1.00
1	1.70	0.86
2	1.80	0.77
3	1.83	0.66
4	2.00	0.59
5	2.07	0.53
6	2.14	0.47
7	2.19	0.42
8	2.22	0.40
9	2.28	0.35
10	2.32	0.31
12	2.40	0.24
14	2.45	0.19
16	2.50	0.15
18	2.53	0.12
20	2.56	0.10
30	2.64	0.03

Source: Courtesy of Bo and Associates Inc.

$$\text{For constant head} \qquad k = q_\infty / FH_c \qquad (4.9)$$

$$C = \frac{q_\infty^2}{\pi n^2} r^2 \qquad (4.10)$$

where
   k is the permeability of soil;
   C is the coefficient of consolidation or swelling;
   q_∞ is the steady state of flow;
   F is the intake factor;
   H_c is the constant head;

Table 4.3   Field hydraulic conductivity test data from a rising head test.

Slug Test Result	Project No.	HG-20201021
Conducted by		Moe. S
Test Location		MW1
Test Date		6-Feb-20
Testing Depth (m)		4.9–7.4 m
Test Formation		Silty Sand
Initial Ground Water Depth, H (m)		1.25
Starting Water Depth, H_0 (m)		3.26

Time Interval (min)	Water Level, h (m)	H-h/(H-H_0)
0	3.26	1.00
1	2.90	0.82
2	2.72	0.73
3	2.55	0.65
4	2.42	0.58
5	2.28	0.51
6	2.17	0.46
7	2.08	0.41
8	1.96	0.35
9	1.88	0.31
10	1.80	0.27
12	1.70	0.22
14	1.63	0.19
16	1.54	0.14
18	1.49	0.12
20	1.43	0.09
25	1.37	0.06
30	1.32	0.03
35	1.29	0.02
40	1.25	0.00

Source: Courtesy of Bo and Associates Inc.

r is the radius of a sphere equal in surface area to that of the cylindrical tip;
n is the slope of the q, $1/\sqrt{t}$ graph.

For variable head $k = A / FT$ (4.11)

where
   k is hydraulic conductivity,
   A is the cross-sectional area of borehole, casing, or standpipe,
   F is the shape factor, and
   T is the basic time factor.

Table 4.4   Field hydraulic conductivity test data from a Constant Head Test.

Constant Head Test	Project No.	HG-20201034
Conducted by		M Karen
Test Location		BH-3
Test Date		3-May-20
Testing Depth (m)		4.5–7.5 m
Test Formation		Silty Sand
Water Level, H (m)		2.67
Radius of well (cm)		2.5
Cross sectional area of well (cm²)		19.6

Time Interval (min)	Volume (cm³)	q (cm³/min)	$1/\sqrt(t)$ (min$^{-\frac{1}{2}}$)
0	0.00		
1	3.14	3.410	1.00
2	1.96	2.500	0.71
3	2.36	2.090	0.58
4	1.57	1.852	0.50
5	1.37	1.685	0.45
6	1.37	1.560	0.41
7	0.98	1.459	0.38
8	0.59	1.385	0.35
9	1.18	1.325	0.33
10	0.79	1.264	0.32
12	1.57	1.190	0.29
14	0.98	1.114	0.27
16	0.98	1.065	0.25
18	0.59	1.010	0.24
20	0.59	0.980	0.22
30	1.47	0.860	0.18
		0.312	0.00

Source: Courtesy of Bo and Associates Inc.

In order to solve the above equation, a shape factor, which varies depending upon the diameter of the hole, and a water intake section are required. The following shape factor is commonly used to solve the above equation:

$$F/D = 2.3\pi \left(\frac{L}{D}\right)\Bigg/\left(log_e\left(1.1\frac{L}{D}\right)\right) + \left(\sqrt{1+1-\left(\frac{L}{D}\right)^2}\right) \qquad (4.12)$$

where
   D is the diameter of the intake section and
   L is the length of the intake section.

Packer tests usually use packer equipment to seal the formation beyond the test section of the rock formation. There are two types of packers available, such

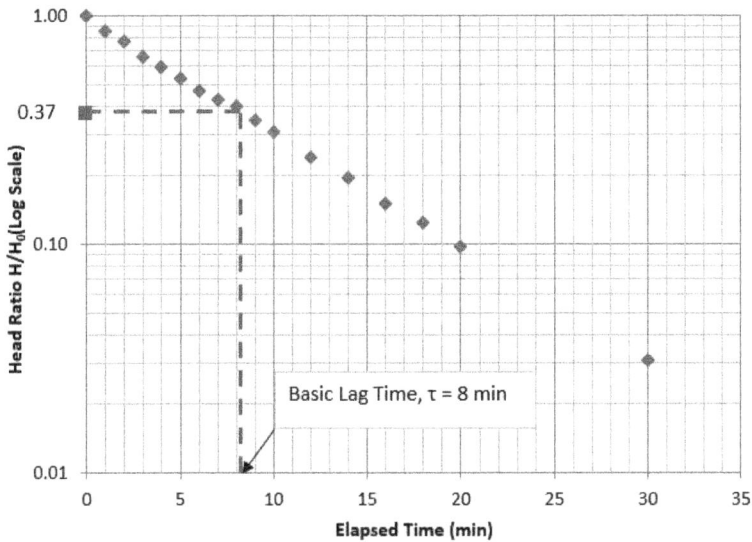

Figure 4.36   Interpretation of hydraulic conductivity test data.

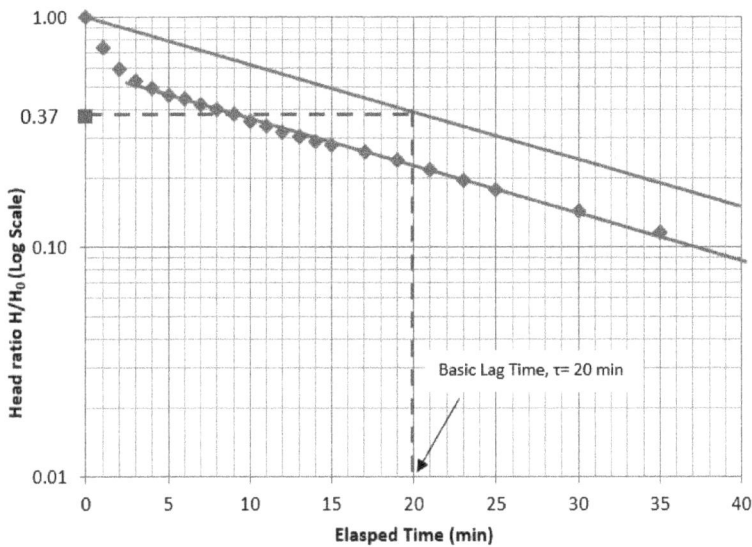

Figure 4.37   Interpretation of hydraulic conductivity test data which required correction for initial drawdown.

as single and double packers. A single packer is usually used when the bottom of the borehole is not required to be sealed if no fracture is expected at the bottom of hole. It is common practice to use a double packer to seal both the top and bottom of the borehole. Figures 4.39 and 4.40 show conceptual drawings of the single and double packer apparatus, respectively.

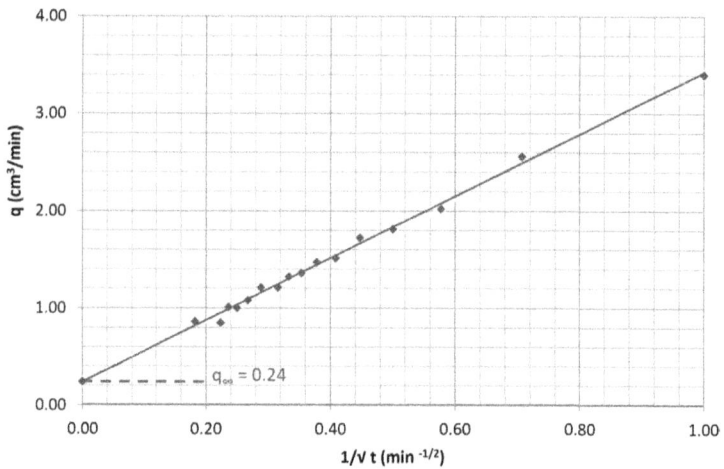

Figure 4.38 Interpretation of constant head test analysis by Gibson's Method.

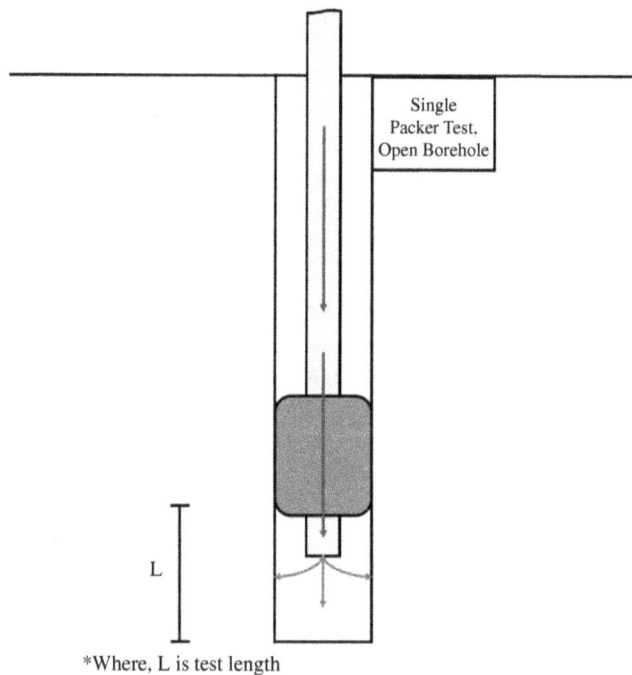

*Where, L is test length

Figure 4.39 Conceptual drawing of Single Packer Test.

After inserting either single or double packer equipment (Figure 4.41) into the required test section using suitable arrangement based on the type of formation testing as shown in the Figures 4.39 and 4.40, the packers are inflated with the help of water pressure using a water pump or air pressure using a compressor to seal the top and bottom or top only of the hole. During the test, water is pumped into the formation through the test section with a gradual increase of specified pressure. The resulting measurements of pressure are interpreted to obtain characteristics of formation and lugeon values. One lugeon is defined as 1 L/min of water flow under 100 meter head of water (Lugeon 1933). Lugeon does not specify the diameter of the borehole, but usually uses a 76-mm-diameter borehole despite the fact that the diameter has limited effect on the test results unless the length of

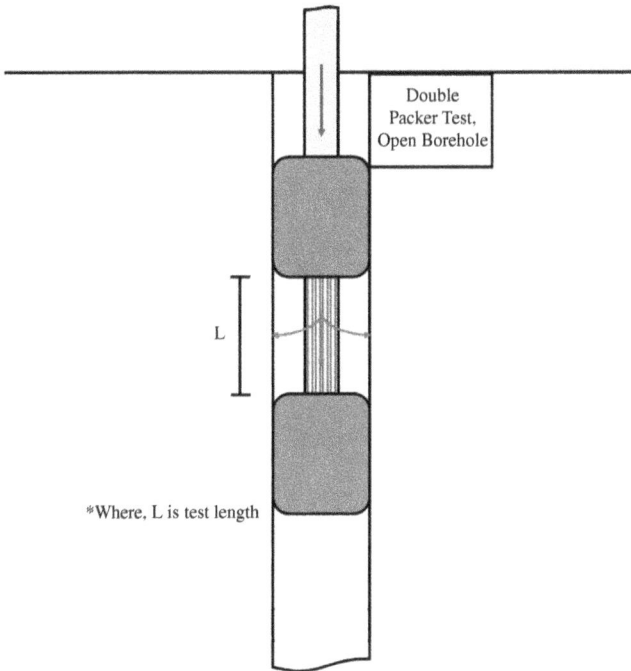

Figure 4.40   Conceptual drawing of Double Packer Test.

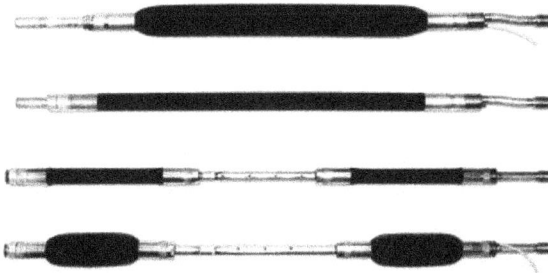

Figure 4.41   Single and double packer equipment.
Source: Courtesy of "Petrometalic".

the test section is too small. From lugeon values, hydraulic conductivity k could be interpreted applying the following equation:

$$k = Q/(2\pi HL) * \log_e L/r \tag{4.13}$$

where
   Q is the rate of injection in cubic meter per sec (m³/s),
   H is the pressure head of water in the test section in meters (m),
   L is the length of the test section in meters (m), and
   r is the radius of the test section in meters (m).

Chapter 5

Specialized *In-situ* Tests and Performance Verification Tests

In addition to boreholes and conventional *in-situ* tests, such as SPT and FVT, many complex geotechnical projects may call for specialized geotechnical *in-situ* testing to characterize the geotechnical properties and parameters of sub-surface formation to be used as input data in the design process. Na *et al.* (1999); Arulrajah *et al.* (2006b, 2011), and Bo *et al.* (2017) have extensively described applications of specialized *in-situ* testing to obtain geotechnical parameters. The following sections describe some typically used specialized *in-situ* testing in detail.

5.1 Cone Penetration Test

Cone Penetration Tests (CPTs) are increasingly popular due to their simplicity and ability to obtain immediate results. There are several types of cones with various functions and capacities manufactured by various manufacturers.

Table 5.1 shows types of cones with various capacities available in the market produced by Geomil. Figures 5.1 and 5.2 show the geometry and design of the Gouda cone. Basic testing procedures involve a continuous penetration of the cone into the sub-soils with a standard penetration rate of 20 mm per second and the recording of the cone resistance (q_c), the sleeve friction (f_s), and the penetration pore pressure (u_{bt}). The CPT can provide several measurements such as cone resistance (q_c), friction (f_s), pore pressure (u_{bt}), and inclination. The CPT can classify the type of soil applying the Robertson and Campanella (1983) chart, (Figure 5.3). Figure 5.4 shows a comparison of the soil profile interpreted from the CPT and observed from the borehole. It can be seen that the CPT can classify the types of soil with reasonable accuracy. Various applications of Cone Penetration Tests were extensively discussed by Bo *et al.* (2001). Some CPT rigs are mounted on trailers, whereas others are mounted on crawlers or trucks as shown in Figure 5.5.

Table 5.1   Types of cones with various capacities available in the market produced by Geomil.

Type of Cone	Max. Cone Value in MN/m²	Max. Friction Value in kN/m²	Max. Charge on Cone in N	Read out on cal. Unit pos. 100 kN	Max. Load on Friction Sleeve in N	Read Out on cal. Unit pos. 20 kN.
25.25	25	250	25.00	500	3.75	375
50.50	50	500	50.00	1.00	7.50	750
75.75	75	750	75.00	1.50	11.25	1.125
100.10	100	1000	100.00	2.00	15.00	1.50

Source: Courtesy of Geomil Equipment B.V., Headquartered Moordrecht, The Netherlands.

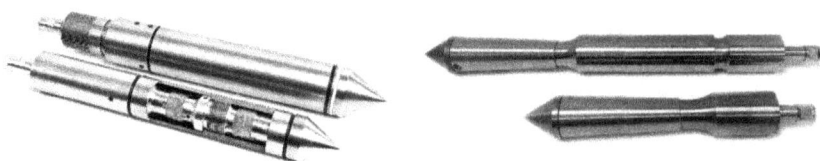

Figure 5.1   Typical cone apparatus used for CPT.

Source: Courtesy of Geomil Equipment B.V., Headquartered Moordrecht, The Netherlands.

There are some CPTs which can be carried out at foreshore locations. Such CPTs are usually operated on the seabed and controlled from a vessel or barge by a remote control (Figures 5.6 and 5.7). CPT tests are carried out in accordance with ASTM D5778-12 and BS 5930-1999.

Several geotechnical parameters can be interpreted using measured parameters from the CPT. Bo *et al.* (2000, 2012, 2017), Arulrajah *et al.* (2004b), and Chang *et al.* (1997) have extensively described applications of Cone Penetration Tests in the characterization of soft clay.

5.1.1 *Undrained shear strength from CPTs*

The undrained shear strength (s_u) can be evaluated from cone resistance using the following equation:

$$s_u = \left(\frac{q_c - \sigma_v}{N_k} \right)$$

(5.1)

where
σ_v is the overburden stress,
N_k is the cone factor.

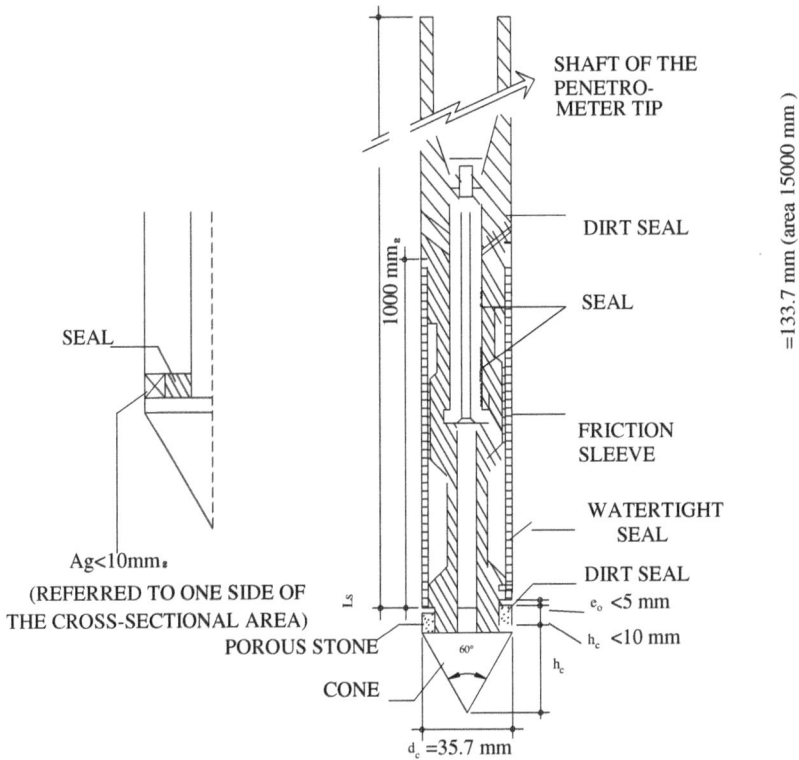

Figure 5.2   Geometry and design of a Gouda piezocone.

Source: Courtesy of Geomil Equipment B.V., Headquartered Moordrecht, The Netherlands.

N_k is reported to be between 11 and 19 based on correction of the field vane shear strength (Lunne and Kleven 1981) and 17 based on a triaxial compression test on non-fissured overconsolidated clay. Kjekstad *et al.* (1978) and Battaglio *et al.* (1986) reported $N_k = 14$ for soft homogenous highly structured $CaCO_3$ Cemented Fucino Lacustrine Clay based on a field vane and triaxial test.

In many cases, corrected cone resistance (q_t) is used in correlating undrained shear strength with cone resistance:

$$s_u = (q_t - \sigma_v) / N_{kt} \qquad (5.2)$$

where

q_t is the corrected cone resistance, and it could be calculated from the cone resistance (q_c) using following equation:

$$q_t = q_c + (1-a)u_{bt} \qquad (5.3)$$

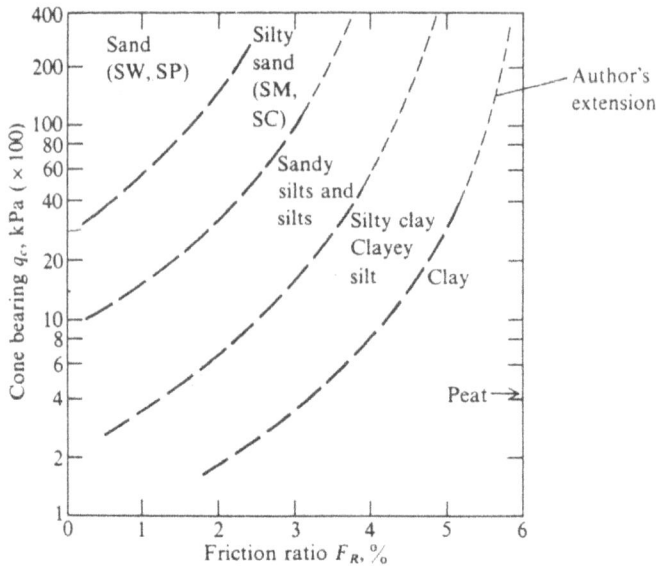

Figure 5.3   Soil classification chart for standard electric cone.
Source: Robertson and Campanella (1983).

where
   a is the unequal area ratio,
   u_{bt} is the pore pressure at the cone base.
   N_{kt} values are reported to be 10–15 for normally consolidated clay and 15–20 for overconsolidated clay (De Ruiter 1982). Dobie (1988) reported N_{kt} values of between 15 and 21 for on-land Singapore Marine Clay. La Rochelle *et al.* (1988); Rad and Lunne (1988) and Powell and Quarterman (1988), reported N_{kt} values between 8 and 29 depending upon Ip based on triaxial compression tests. Aas *et al.* (1986) proposed a relationship between N_{kt} and plasticity index (Ip) in clay as follows:

$$N_{kt} = 13 + \left(\frac{5.5}{50} \right) I_p \left(\pm 2 \right) \tag{5.4}$$

Bo *et al.* (2000) has reported the relationship between N_{kt} and Singapore Marine Clay as follows:

$$N_{kt} = 23.8 - \left(0.263 I_p \right) \tag{5.5}$$

CONE RESISTANCE (MN/m2)

LOCAL FRICTION (kN/m2)

PORE PRESSURE (Bar)

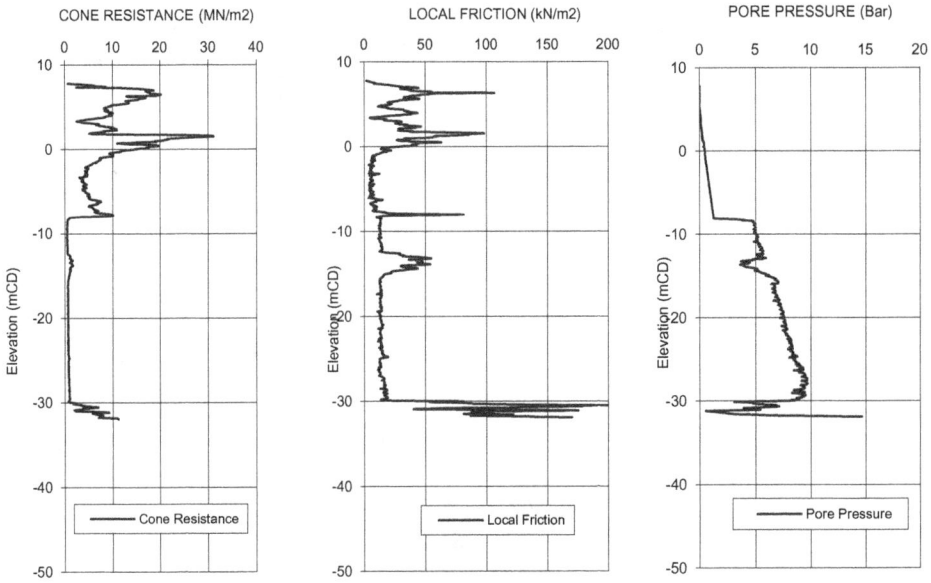

A4 Area
SOIL PROFILE (PC-200)

A4 Area
SOIL PROFILE (DC-961)

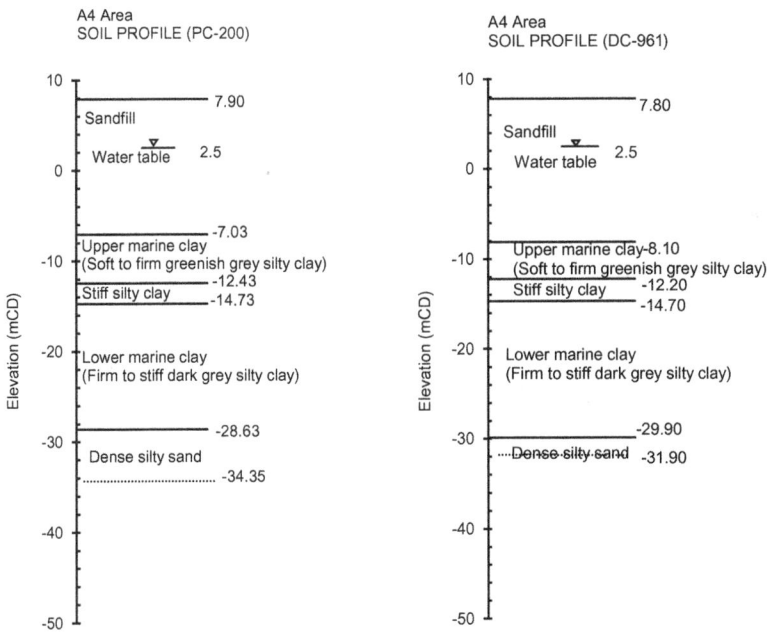

Figure 5.4   Comparison of soil profile interpreted from CPT and borehole.

Source: Bo *et al.* (2019).

Figure 5.5   Typical CPT Rig; left, mounted on truck; interior and exterior.
Source: Courtesy of Geomil Equipment B.V., Headquartered Moordrecht, The Netherlands.

By using the above correlation, the undrained shear strength of clay can be well predicted from q_c values. The comparison of predicted and measured field vane shear strengths is shown in Figure 5.8 for Singapore Marine Clay at the study area.

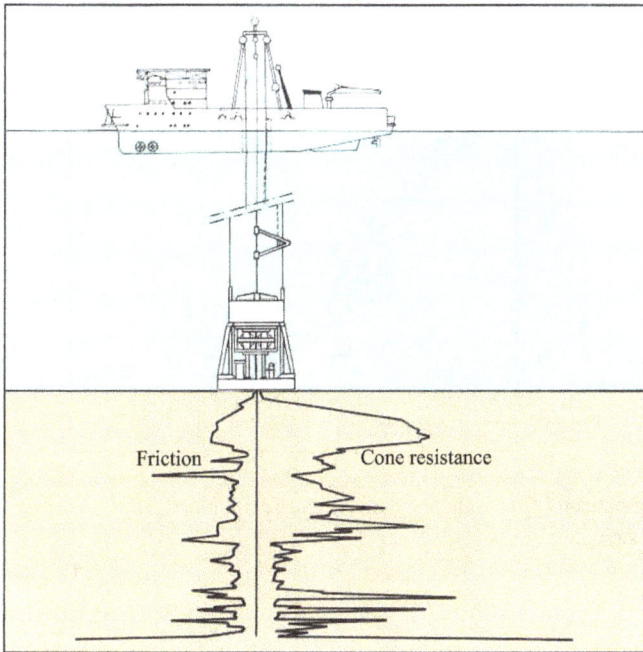

Figure 5.6 Typical process of CPT on seabed.
Source: Bo and Choa (2004).

Figure 5.7 CPT operated on the seabed and controlled with remote control from a vessel or barge.
Source: Bo and Choa (2004).

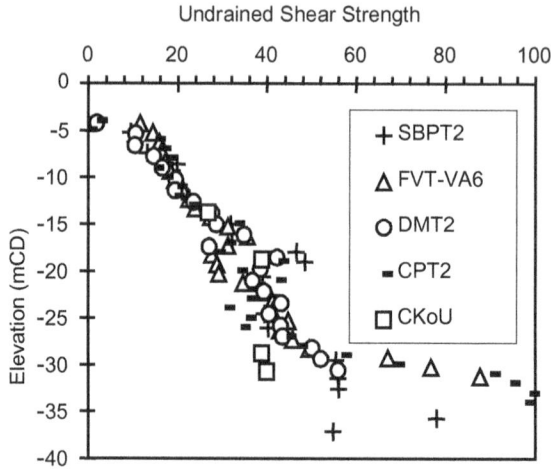

Figure 5.8   Comparison of undrained shear strengths determined from specialist *in-situ* tests and field vane shear strength tests.

Several others have reported various N_{kt} values for various clays from all over the world either based on triaxial tests or field vane test data. Some of those reported values, based on triaxial tests, are N_{kt} of 13.7 for Newcastle clay in Australia (Jones 1995), N_{kt} between 13.5 and 15.5 for Sarapui soft clay in Brazil (Rocha-Filho and Alencar 1985), between 10.3 and 15 for Recife soft clay in Brazil by Coutinho *et al.* (1993), and N_{kt} ranging between 12 and 20 for normally consolidated clay in the Southern part of Nigeria by George and Ajayi (1995).

Reported N_{kt} values based on field vane tests are $N_{kt} = 14.5$ for Jacarepaqua clay in Brazil by Rocha-Filho (1987), $N_{kt} = 15$ for Porto Alegre soft clay by Soares *et al.* (1986) and Quilombo soft clay by Arabe (1995), $N_{kt} = 10$ for different deposits of clay in Denmark by Denver (1988), Kammer Mortensen *et al.* (1991) and Jorgensen and Denver (1992), and N_{kt} of 9 to 14 for Japanese marine clay by Tanaka (1994). Tanaka also reported N_{kt} values of between 8 and 16 based on laboratory Unconfined Compression Test.

In Germany, deduction of overburden pressure is not taken into account and the cone factor is also not used. A direct relationship between cone resistance (q_c) and undrained shear strength was proposed as follows:

$$S_u = \frac{q_c}{N} \tag{5.6}$$

where
    N is varied between 10 and 20.

$N = 12$ for soft clay and 20 for OC clay was reported by Cao (1997). Sanglerat (1972) reported $N = 10$ for $q_c < 0.5$ MPa and $N = 18$ for $q_c > 0.5$ MPa. A similar direct correlation was also used in Vietnam and reported Nc values were 20 for soft silty clay in Vietnam (Nhuan *et al.* 1985).

5.1.2 *Overconsolidation ratio from CPTs*

In addition to undrained shear strength, the CPT is also a useful to predict the overconsolidation ratio (OCR) of soft clay. The OCR can be estimated from q_t using the following equation:

$$OCR = \alpha \left[\frac{q_t - \sigma_{vo}}{\sigma'_{va}} \right] \tag{5.7}$$

where σ'_{vo} is the effective overburden stress and α is constant, ranging from 0.2 to 0.5. The α value of 0.33 was reported based on the CPT pore pressure measured on the shoulder of a cone (Kulhawy and Mayne 1990) and $\alpha = 0.81$ based on the mid-face element (Chen and Mayne 1996). Senneset *et al.* (1982) and Konrad and Law (1987) reported α values of 0.49 based on pore pressure measurements on the shoulder. For Singapore Marine Clay, Bo *et al.* (1998b) proposed the α value of 0.32.

Figure 5.9 shows a comparison of the OCR interpreted from various *in-situ* tests and that interpreted from laboratory oedometer tests.

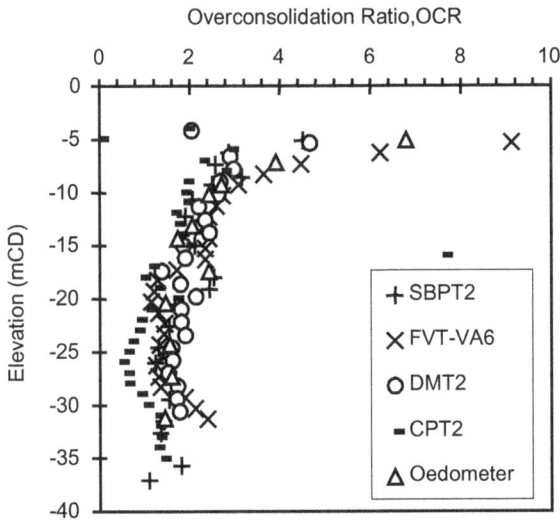

Figure 5.9   Comparison of overconsolidation ratio determined from specialist *in-situ* tests and laboratory measurements.

5.1.3 *Coefficient of consolidation due to horizontal flow (C_h) from CPTs*

Since the CPT equipment has a pore pressure transducer, it is also useful to carry out the pore pressure dissipation test in the clay. When the CPT penetrates into the clay, dynamic pore pressure occurs. However, if the cone is held at the same position for a longer duration, the dynamic pore pressure will dissipate with time. This pore pressure dissipation curve can be analyzed applying the Baligh and Levadoux (1980) strain path method. The coefficient of consolidation due to horizontal flow C_h values can be calculated from relevant time factor T using the following equation:

$$C_h = \frac{R^2 T}{t} \tag{5.8}$$

where
   R is the radius of the pushing cone in meters,
   T is a dimensionless time factor,
   t is the time elapsed to reach a given degree of consolidation in years.

The resulting C_h values are required to be corrected to the normally consolidated (NC) condition using the recompression ratio. Figure 5.10 shows some pore pressure dissipation curves measured by the CPTU tests and Figure 5.11 shows a comparison of C_h values measured from various types of laboratory and field *in-situ* tests. Bo *et al.* (1998a) and Arulrajah *et al.* (2009) have extensively described several field dissipation tests on Singapore marine clay.

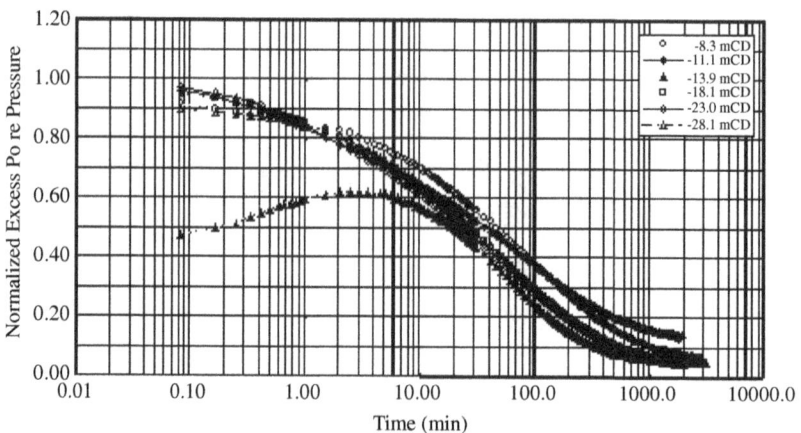

Figure 5.10   Pore pressure dissipation curves measured by CPTU test.

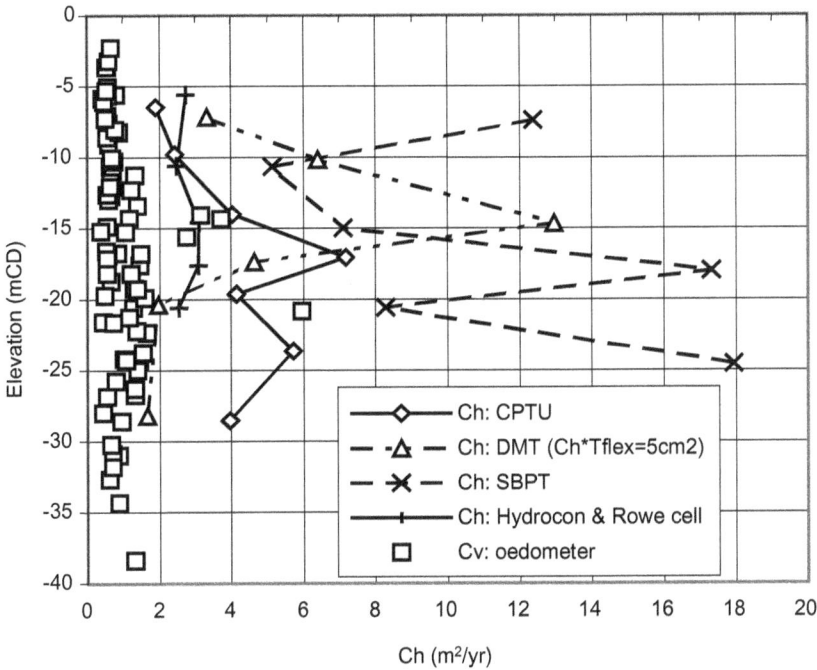

Figure 5.11   Coefficient of consolidation due to horizontal flow.
Source: Bo *et al.* (2003a) and Arul *et al.* (2004b).

5.2 Flat Dilatometer Test (DMT)

A Marchetti flat dilatometer (Marchetti 1980) with a steel membrane on one side of the blade (Figure 5.12) is a specialized *in-situ* testing equipment to characterize geotechnical parameters of the soils. The test involves pushing the flat dilatometer into the seabed with a 20 ton static rig at a standard penetration rate of 20 mm per second. The pushing is to be temporarily stopped at selected depth intervals in order to record two pressure readings corresponding to two prefixed states of expansion of the membrane. The first reading corresponds to the membrane lift-off pressure and the second reading to the pressure required for the center of the membrane to deflect by a pre-set distance of 1mm into the soil. These readings, called P_0 and P_1, respectively, are taken after allowing for effect of the membrane stiffness. The testing procedure follows the instructions in the dilatometer operation manual prepared by Marchetti and Crapps (1981). Some details for the procedure can also be found in ASTM D 6635-15. Figure 5.13 shows photographic features of the flat dilatometer blade and flat dilatometer testing in progress.

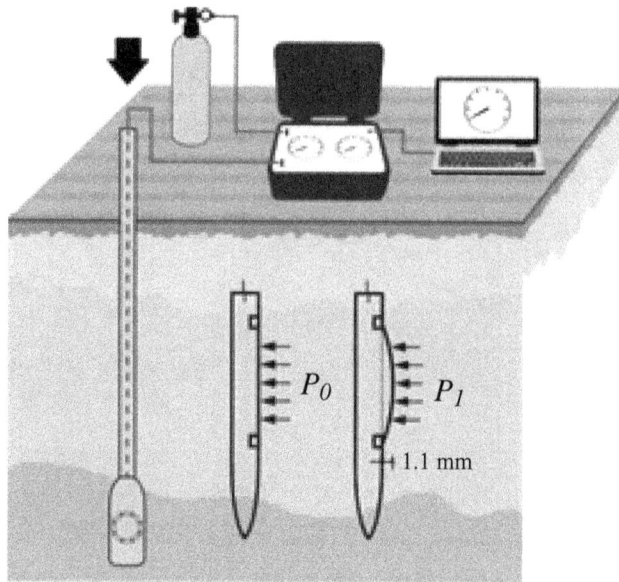

Figure 5.12   Marchetti flat dilatometer blade with steel membrane on one side.
Source: Courtesy of Marchetti..

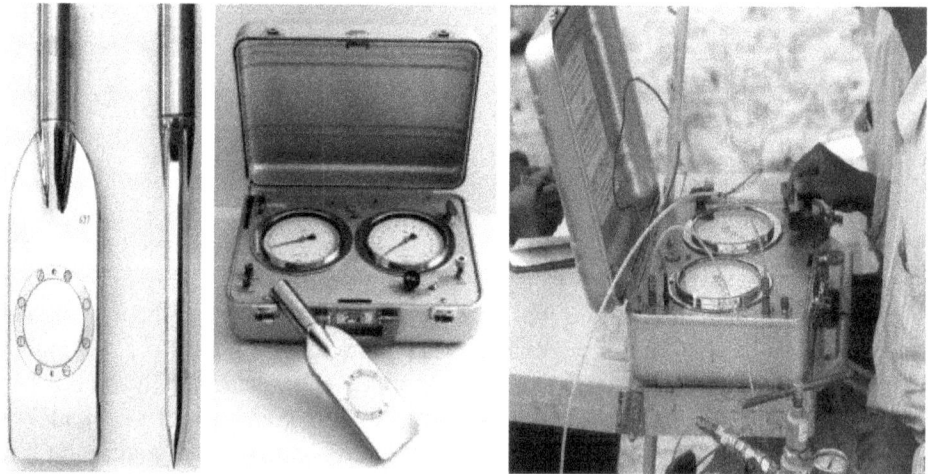

Figure 5.13   Photographic features of flat dilatometer blade and flat dilatometer testing in progress.
Source: Bo and Choa (2004).

The flat dilatometer measures two pressure values called P_0 and P_1. For these two values, three indices such as material Index (I_D), horizontal stress index (K_D), and dilatometer modulus (E_D) could be obtained using the following equations:

$$I_D = \frac{(P_0 - P_1)}{(P_0 - u_0)} \qquad (5.9)$$

$$K_D = \frac{(P_0 - u_0)}{(\sigma_{v0} - u_0)} \qquad (5.10)$$

$$E_D = 34.7(P_1 - P_0) \qquad (5.11)$$

where
  u_0 is the pre-insertion pore water pressure.
  Marchetti (1980) proposed the classification of soil using material index values.
  Figure 5.14 shows measured and calculated indices from DMT tests for a test area. It was found that the dilatometer could classify the soil type accurately.

5.2.1 *Undrained shear strength from DMTs*

Like the CPT, s_u can also be estimated from K_D values obtained from the DMT. Marchetti (1980) proposed undrained shear strength s_u with a lateral stress index (K_D) as follows:

$$s_u = 0.22\sigma'_{v0}\left(\frac{K_D}{2}\right)^{1.25} \qquad (5.12)$$

  Bo *et al.* (2000) proposed a power function of 1.0 for upper and intermediate Singapore Marine Clay and 0.7 for lower Singapore Marine Clay instead of 1.25. Figure 5.8 shows the s_u values estimated from the DMT.

5.2.2 *Overconsolidation ratio from DMTs*

From the lateral stress index K_D, the OCR of clay can be estimated as proposed by Marchetti (1980):

$$OCR = 0.5K_D^{1.56} \qquad (5.13)$$

  Bo *et al.* (1998b) proposed the power function of 1.0 for lower and upper Singapore Marine Clay and 0.8 for intermediate Singapore Marine Clay. Figure 5.9 also shows the OCR estimated from the DMT.

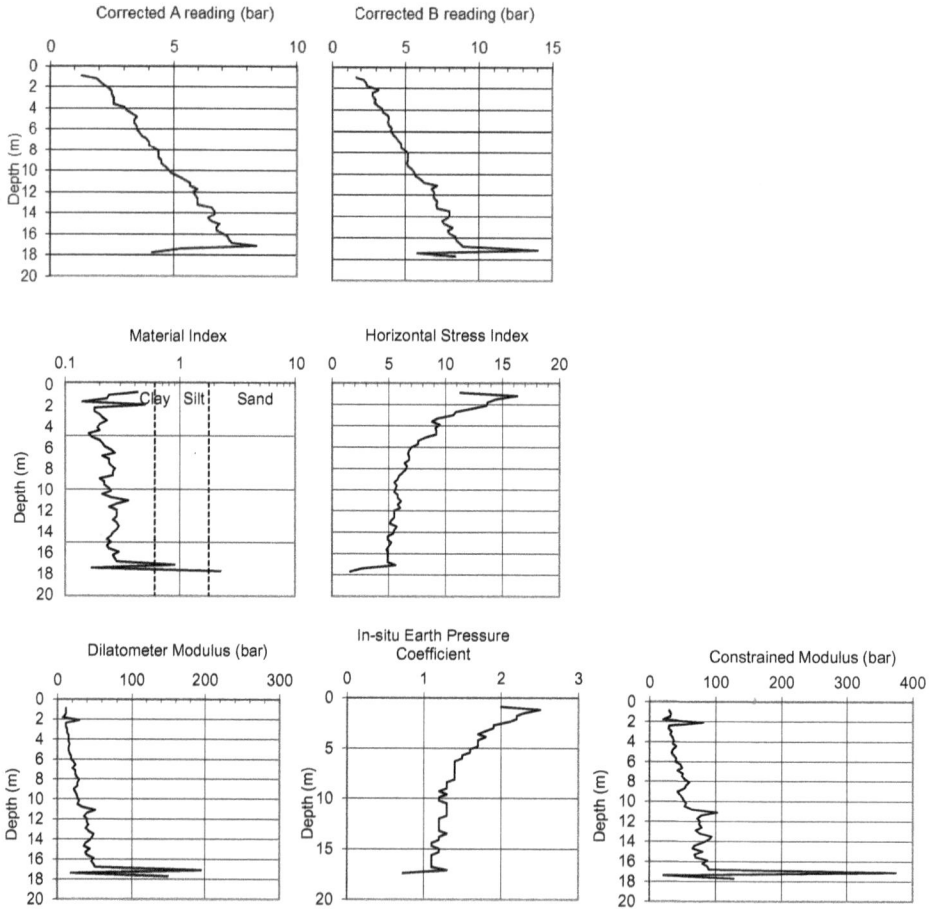

Figure 5.14   Measured and calculated indices from DMTs.

5.2.3 *Coefficient of consolidation due to horizontal flow (C$_h$) from DMTs*

DMT test measures the total stress of the soil. Therefore from dissipation of total lateral stresses, C_h values again can be calculated using the following equation proposed by Marchetti and Totani (1989):

$$C_{h_{(DMTA)}} \, T_{flex} = 5 - 10 \; cm^2$$

(5.14)

where

T_{flex} is the dimensionless time factor.

Figure 5.15 shows the dissipation curve from DMTs for upper Marine Clay.

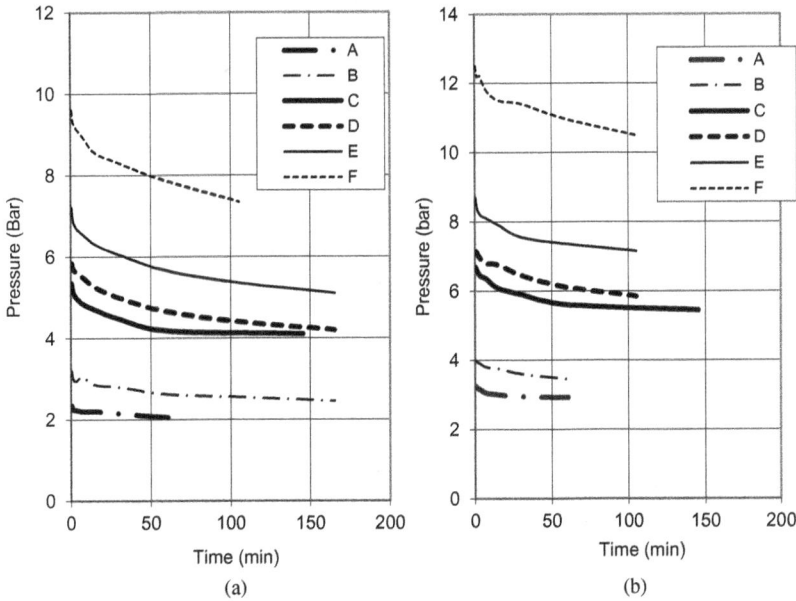

Figure 5.15   (a) A Reading dissipation curve from DMT; (b) C reading dissipation curve from DMT (right). Legend: A – 3.6 & 3.9 depth, B – 6.6 & 6.9 depth, C – 9.3 & 9.6 depth, D – 14.7 & 15 depth, E – 19.5 & 19.8 depth, and F – 25.5 & 25.8 depth.

From the C reading dissipation test, C_h is given by

$$C_{h_{(DMTC)}} = 600 \left[\frac{T_{50}}{t_{50}} \right] mm^2/ \, min \qquad (5.15)$$

Figure 5.15 shows C reading dissipation curves in upper marine clay. Figure 5.11 also shows C_h values interpreted from DMT dissipation tests compared with those from other type of *in-situ* and laboratory tests.

DMTs also can be carried out in the offshore regions using a seafloor dilatometer. The Seafloor Dilatometer (Seafloor DMT) was developed to perform DMTs operating directly from the seabed (Figure 5.16). The machine is composed of a pushing unit with an approximate weight of 60-80 kg, and it is easily transportable.

The Seafloor DMT was designed to operate up to a water depth of about 100 m and is able to apply up to a 5 ton push. Usually, six or seven penetration rods are pre-charged vertically on top of the pushing system before lowering the machine to the seabed. Additional rods may be added as long as verticality in the rod string is ensured.

Figure 5.16   A Seafloor Dilatometer testing in progress.
Source: Courtesy of Marchetti.

Considering that penetration speed does not influence DMT readings, which are taken when penetration is stopped, the Seafloor DMT was designed to push with multiple short length strokes (ex. 0.10 m). Figure 5.16 shows seafloor Dilatometer testing in progress.

DMTs also utilize an automated dilatometer called the Medusa DMT, which is able to autonomously perform dilatometer tests (Marchetti *et al.* 2019; Marchetti 2019). An electronic board, powered with rechargeable batteries, activates a motorized syringe for hydraulically expanding the DMT membrane (Figure 5.17). The blade has the same dimensions as the original standard flat plate dilatometer. The device may operate in a cableless manner (MEMO mode), a valid option especially in offshore projects at medium depths to depths greater than 100 m. Whenever possible, the Medusa is operated with an electric cable, to obtain real-time results during test execution.

The Medusa DMT is capable of measuring the total horizontal pressure of the soil with time, suggesting some potentiality for improving K_0 and OCR interpretation in sand, for characterizing partially draining soils (Schnaid *et al.* 2018), and for extending the range of soils to perform dissipation tests for estimations of consolidation and permeability coefficients.

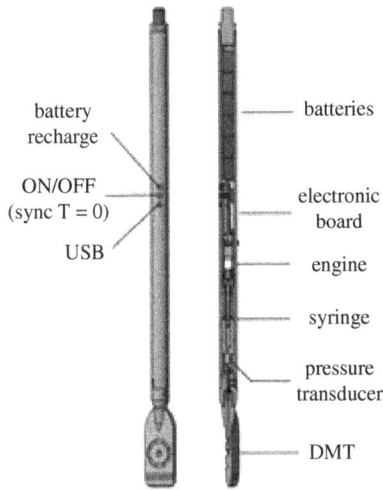

Figure 5.17   The properties of Medusa DMT.

Source: Courtesy of Marchetti.

5.3 Self-boring Pressuremeter Test (SBPT)

A Cambridge-type self-boring pressuremeter (Wroth 1984) with strain-measuring arms located at the mid-level, as shown in Figure 5.19, is used in *in-situ* tests to characterize the soils. The instrument has strain gauge type transducers attached to the center core or pressuremeter body, which is covered with a rubber membrane, for direct recording of the radial displacement and the applied pressure.

The self-boring pressuremeter is equipped with a rotary bit at the base. The SBPT involves, at first, advancement and insertion of the pressuremeter into the selected depth in the ground using the self-boring technique. Following the insertion of the apparatus, the rubber membrane is inflated by the injection of gas pressure and both the applied pressure and the corresponding displacement of borehole (cavity) wall are measured. The test procedure generally follows Mair and Wood (1987) and Hawkins *et al.* (1990). The test results are usually presented in a plot of applied pressure versus (radial) cavity strain, which can be interpreted by the cavity expansion theory. Figure 5.18 shows typical test results from the self-boring pressuremeter test. Figure 5.19 shows the geometry and dimensions of the self-boring pressuremeter and Figure 5.20 shows self-boring pressuremeter testing in progress.

Tests can be carried out either by stress control or strain control. The test procedure for pressuremeter testing can be found in ASTM 4719-20 and BS 5930-1999. From the test, the following basic measurements can be obtained:

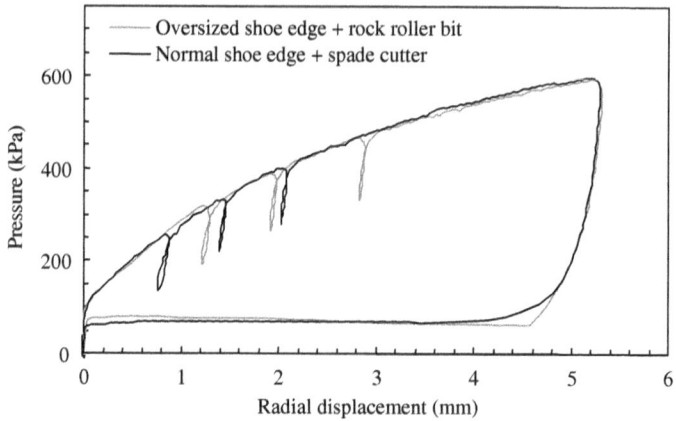

Figure 5.18   Typical test results from self-boring pressuremeter test.
Source: Bo and Choa (2004).

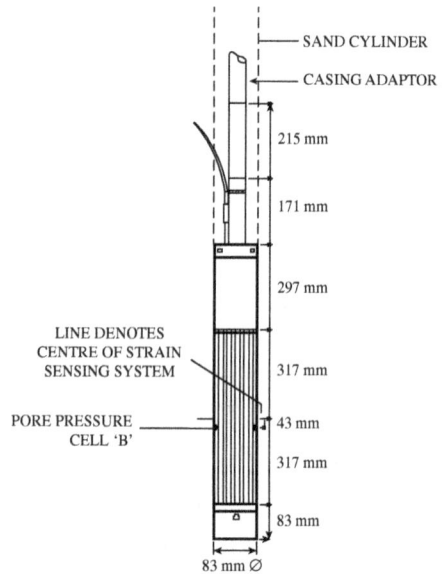

Figure 5.19   Geometry and dimensions of a self-boring pressuremeter.
Source: Bo and Choa (2004).

- Lift-off pressure
- Stress vs strain curve
- Several unload, reload loop
- Limit pressure (PL)
- Pore pressure dissipation curve from dissipation test.

Figure 5.20   Self-boring pressuremeter testing in progress.
Source: Bo and Choa (2004).

From the lift-off pressure, the lateral earth pressure can be obtained. From the stress–strain curves, various types of moduli such as initial tangent moduli, secant moduli, and unload–reload moduli can be obtained.

5.3.1 *Undrained shear strength from SBPTs*

Undrained shear strength can also be estimated from limit pressure obtained during SBPTs by using the following equation:

$$S_u = \frac{(PL - \sigma_{h0})}{N_p} \qquad \text{where } N_p = 1 + \log_e\left(\frac{G}{C_u}\right) \qquad (5.16)$$

where PL is limit pressure,
   σ_{h0} is total horizontal stress,

$\dfrac{G}{C_u}$ is shear modulus, and

N_p is pressuremeter constant.

Marsland and Randolf (1977) adopted N_p values ranging between 5.5 and 6.8. It could be again suggested that the N_p values for specific clays should be locally obtained by empirical correlation. Bo *et al.* (2000) suggested that N_p values for Singapore Marine Clay at Changi are 6.0, 6.4, and 7.2 for upper, intermediate, and lower marine clays, respectively. Figure 5.8 also shows a comparison between field vane shear strength and that interpreted from SBPTs.

5.3.2 *Overconsolidation ratio from SBPTs*

Since the self-boring pressuremeter can measure the total horizontal stress, it is possible to determine the K_0 values, hence the OCR can be estimated. Figure 5.9 also shows the OCR interpreted from the self-boring pressuremeter compared with laboratory results:

$$OCR = \left[\frac{K_{0oc}}{K_{0nc}}\right]^{1/h}$$

(5.17)

where $h = 0.32$–0.40,

K_{0oc} is earth pressure coefficient at rest under overconsolidated conditions, and
K_{0nc} is earth pressure coefficient at rest under normally consolidated conditions.

5.3.3 *Coefficient of consolidation due to horizontal flow (C_h) from SBPTs*

From the pore pressure dissipation test carried out using SBPTs, the coefficient of consolidation due to horizontal flow (C_h) can be estimated using the following equation:

$$C_h = \frac{T_{50}\rho^2}{t_{50}}$$

(5.18)

where t_{50} is the time taken for the excess pore pressure to fall to half of its maximum value. T_{50} is the time factor and ρ is the radius of the cavity.

k_h can be calculated from C_h values. k_h interpreted from various *in-situ* tests are shown in Figure 5.21. Interpretation of k_h from various *in-situ* tests can be found in Bo *et al.* (1998a).

5.4 Cone Pressuremeter Test (CPMT)

A cone pressuremeter is a combination of a cone penetrometer and a pressuremeter. Therefore, it can measure the parameters measured by both the CPT and a pressuremeter. However, the CPT cone is usually bigger than the conventional cone and its cone base area is usually 15 cm². A CPT is carried out in the same manner as a standard size CPT and measured the same parameters which can be obtained from standard CPTs. A pressuremeter is attached above the cone and its diameter is 43.7 mm and length 2 meter.

A test can be carried out in the same manner as SBPTs and measure the same parameters; hence, the same sets of geotechnical parameters can be obtained like

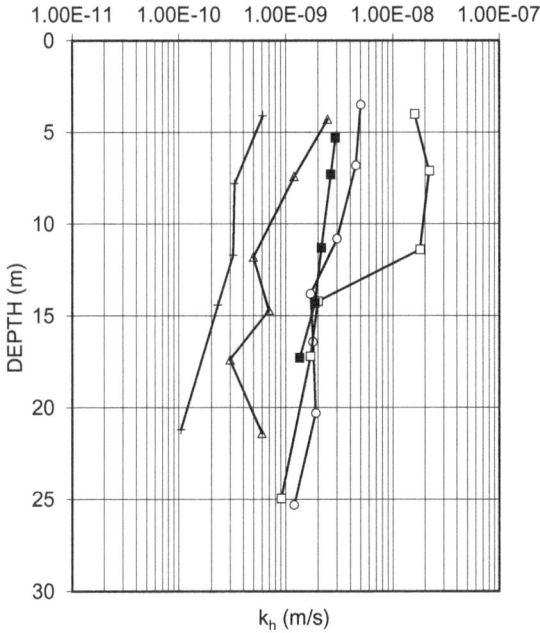

Figure 5.21 k_h interpreted from various *in-situ* tests.
Source: Bo *et al.* (1998a).

SBPTs. Figure 5.22 shows the geometry and dimensions of the cone pressuremeter. The advantage of CPMTs is that pre-boring or self-boring is not required.
However, soil disturbance in the clay and contraction in the granular soil can occur
due to the penetration. However, this type of CPMT is suitable for granular soil,
where maintaining regular-sized boreholes is difficult with the pre-bore method.
Figure 5.23 shows some geotechnical parameters measured by CPMTs prior to
and post ground improvement works.

5.5 Seismic Cone Test (SCPT)

The seismic cone is a combination of a CPT and a seismic geophone receiver. The
CPT can be carried out as conventional CPT, and similar sets of geotechnical
parameters can be collected and interpreted. Figure 5.24 shows the geometry and
dimensions of a seismic cone. At a certain interval, the penetration of the cone can
be paused to carry out the seismic measurements. Normally, seismic force is provided by a hammer to the wooden plate and the seismic wave is detected with a
receiver near the cone tip. Figure 5.25 shows seismic cone testing in progress.
From the measured data, the compression wave (P) and shear wave (S) can be
calculated. From there, compression and shear wave velocity, (ν_p) and (ν_s),
respectively, can be calculated. Figure 5.26 shows shear velocity, shear modulus,

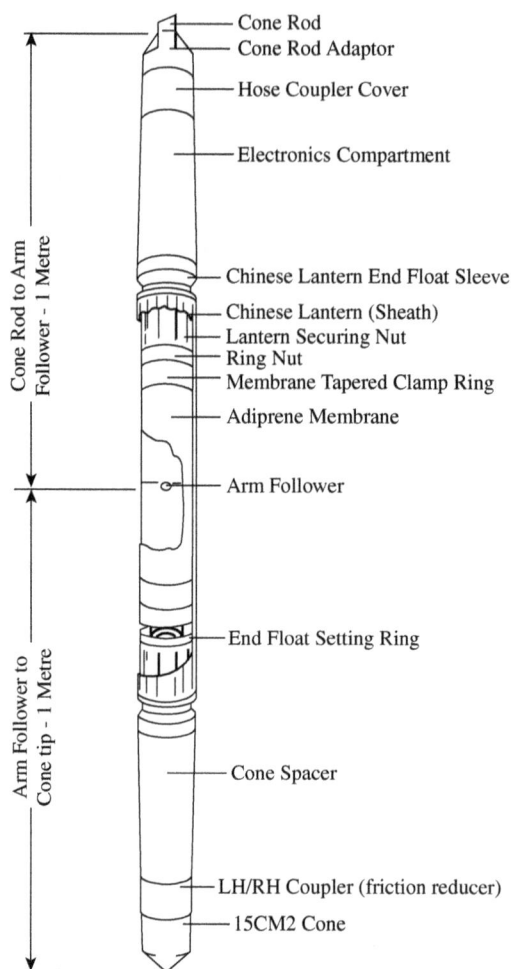

Figure 5.22   Geometry and dimensions of a cone pressuremeter.
Source: Soe Moe *et al.* (2019).

and cone resistance measured from the seismic cone test. Small strain shear modulus (M_0) and constrained modulus (M) can be estimated from ν_s and ν_p using the following formulae:

$$v_o = \rho\left(v_s\right)^2 \tag{5.19}$$

$$M_0 = \rho\left(v_p\right)^2 \tag{5.20}$$

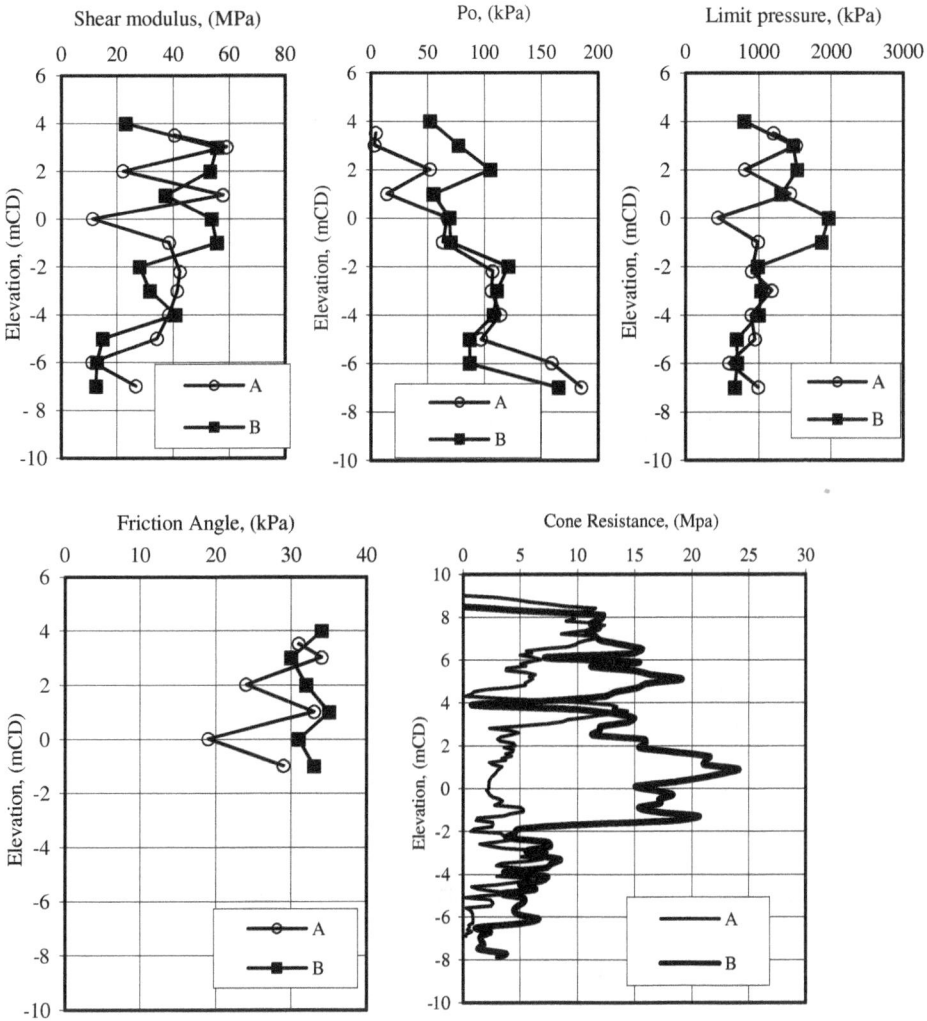

Figure 5.23   Geotechnical parameters measured by CPMT and modulus and cone resistance measured with CPMT before and after ground improvement.

Note: A: CMPT Before Ground Improvement and B: CMPT After Ground Improvement.

5.6 Seismic Dilatometer

The Seismic Dilatometer (SDMT) is a combination of the Flat Dilatometer with an add-on seismic module for measuring the shear wave velocity (Marchetti S. *et al.* 2008) and optionally also the compression wave velocity (Amoroso *et al.* 2016). The seismic module is an instrumented steel rod placed just above the DMT blade and equipped with two receivers spaced 0.5 m. When a

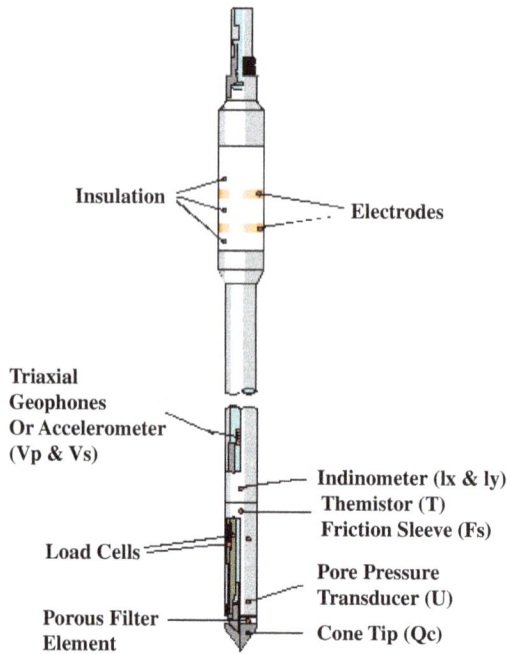

Figure 5.24   Geometry and dimensions of a seismic cone.
Source: Bo and Choa (2004).

Figure 5.25   Seismic cone testing in progress.

shear or compression wave is generated at the surface, it first arrives at the upper receiver, then, after a delay, at the lower receiver. The wave traces of the two receivers are amplified and digitized at depth and transmitted to the computer at the surface. The software processes the signals and evaluates the arrival delay, providing a real-time interpretation of the wave velocity. Figure 5.27 shows that the shear wave velocity *Vs* is obtained as the ratio between the difference of the

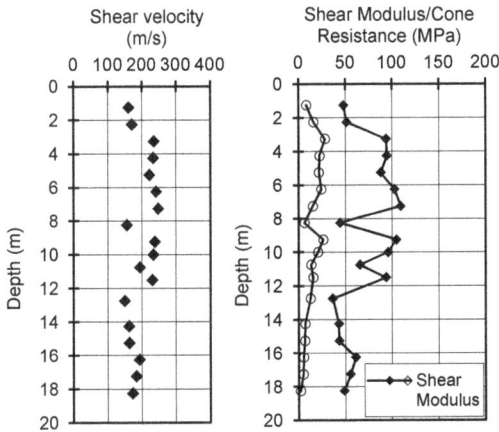

Figure 5.26   Shear and constraint modulus measured from seismic cone test.

wave travel path from the source to the receivers (S2–S1) and the wave arrival delay *t* from the first to the second receiver.

Digital acquisition at depth, combined with a true interval configuration, enables the SDMT to provide high accuracy *Vs* profiles with a repeatability of typically within 1% (i.e., a few m/s). Several comparisons and case histories have shown very good agreement between SDMT and crosshole results in different soil types (Amoroso *et al.* 2015; Décourt *et al.* 2016; Pein *et al.* 2019).

The SDMT may be employed in penetrable soils similar to the DMTs, but also in non-penetrable soils. In this second case, the tests are performed in a sand backfilled borehole.

5.7 Dynamic Cone Testing and Auto Ram Sounding

Dynamic cone testing is usually carried out in coarse granular soils in which SPTs cannot be performed as the coarse granular soil could damage the SPT sampler. Dynamic cone testing can be carried out in the boreholes at the bottom of holes. A dynamic cone can be driven using the same SPT hammer by dropping the hammer from 750 mm height. The cone can be driven continuously without advancing the borehole. The cone usually used with the SPT hammer has a diameter of 50.5 mm and an angle of 90 degree. The testing procedure can be found in ASTM D 6951-03 and BS 5930-1999.

Dynamic cone testing utilizes equipment called Swedish ram sounding, which can carry out automatic dynamic probing with a solid cone. There are several

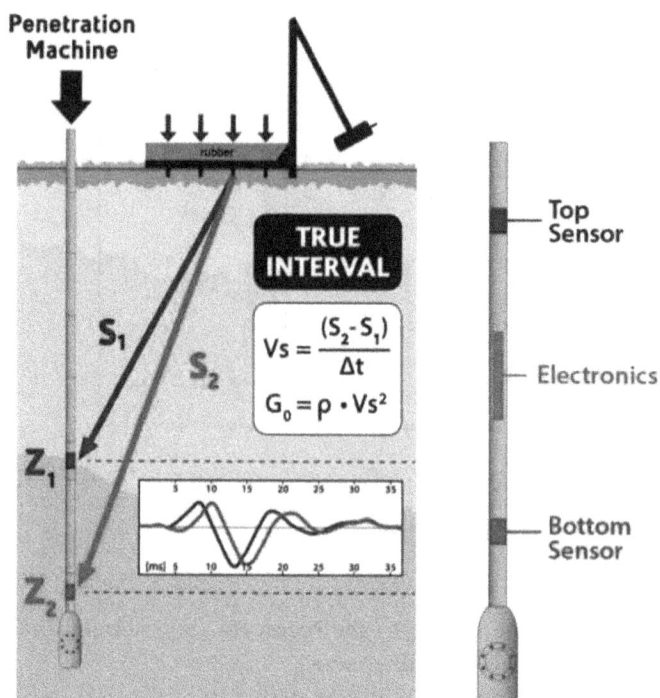

Figure 5.27   Shear Velocity obtained using a seismic dilatometer.
Source: Courtesy of Marchetti D.

types of ram soundings such as light duty, medium, and heavy duty. The hammers used in various categories are shown in Table 5.2. The drop height is usually 50 cm and the number of blows is counted for every 20 cm penetration. The ram sounding can detect the relative density of granular soils. Figure 5.28 shows the geometry and dimensions of auto ram sounding equipment and Figure 5.29 shows auto ram sounding in progress. Ram sounding is useful in quality control of compaction works to verify the relative density improvement. Figure 5.30 shows a comparison of prior and post compaction ram sounding results.

5.8 BAT Permeameter

In addition to the conventional hydraulic conductivity tests described in the previous chapter, there is a probe which can be pushed into the soil at the bottom of a borehole to measure hydraulic conductivity which was developed by Bengt A Torstensson (1984).

The BAT permeameter measures the pore pressure locally in the soil, with little water movement, resulting in a quick reaction time in soft soils

Table 5.2   Various types of Auto ram sounding equipment.

Type	Abbreviation	Mass (kg)	Drop Height (cm)
Light	DPL	≤ 10	50
Medium	DPM	> 10 < 40	20–50
Heavy	DPH	≥ 40 ≤ 60	50
Super Heavy	DPSH	> 60	50

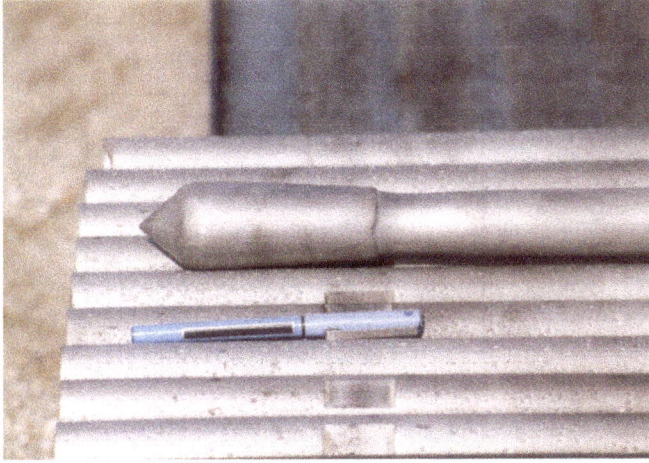

Figure 5.28   Auto ram sounding cone.
Source: Bo and Choa (2004).

Figure 5.29   Auto ram sounding in progress.
Source: Bo and Choa (2004).

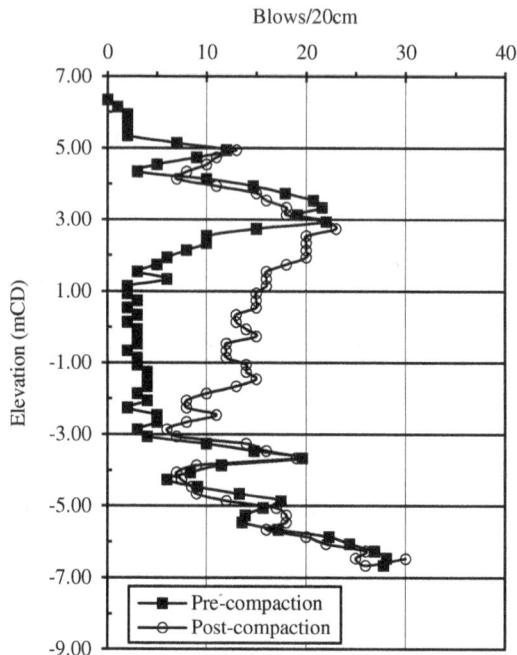

Figure 5.30   Comparison of prior and post auto ram sounding at compaction area.

(Torstensson & Schellingerhout 1999). The BAT permeameter has the ability to sample groundwater, measure pore water pressure, and determine *in-situ* hydraulic conductivity of soft soils.

The key element in the BAT system is the filter tip, which consists of a thermoplastic body and a porous plastic filter tip (Torstensson 1984). The diameter of the BAT filter used is 30 mm and the length is 40 mm. The different test adapters make a tight temporary connection to the filter tip with the aid of a hypodermic needle. The pore pressure adaptor contains a hypodermic needle and an electronic pressure transducer, connected to a battery-operated digital readout unit via a cable. The pore pressure adaptor is threaded to an extension pipe and lowered into a pre-bored borehole and placed at the desired elevation. When the pore pressure adaptor is lowered into the borehole, it is coupled to the nozzle in the filter tip and gravity draws the hypodermic needle downward, penetrating the rubber disc mounted in the filter tip. The needle provides a hydraulic connection between the interior of the filter tip and the test adapter (Torstensson 1984). In the BAT permeameter test, a penetrometer has to be pushed into the clay and causes a smear around the BAT permeameter. This effect affects the k_h measurement, similar to the insertion of a mandrel, and must be considered prior to the measurements (Bo *et al.* 2003a). Figure 5.31 shows the geometry and dimensions of the BAT permeameter.

The *in-situ* measurement of hydraulic conductivity can be carried out either as an inflow test or as an outflow test. In the former case, the gas/water container is completely filled with gas at the start of the test. An inflow test can be conducted simultaneously with the extraction of the pore water sample. In an outflow test, the container is partially filled with compressed gas. The air in the chamber is evacuated (or pressurized) to any desired pressure. As water flow into (or out of) the probe causes a change in air pressure in the chamber, a pressure transducer monitors the pressure change. The test is based on the measurement of flow into and out of a sample container. This rate is computed by the pressure change measured in the container using Boyles's law, which can be translated into a volume change. Analysis of the time-pressure record thus yields the horizontal hydraulic conductivity. The quantity of flow and head is computed from the change in the gas pressure measured in the chamber using Boyle's Law (Torstensson 1984):

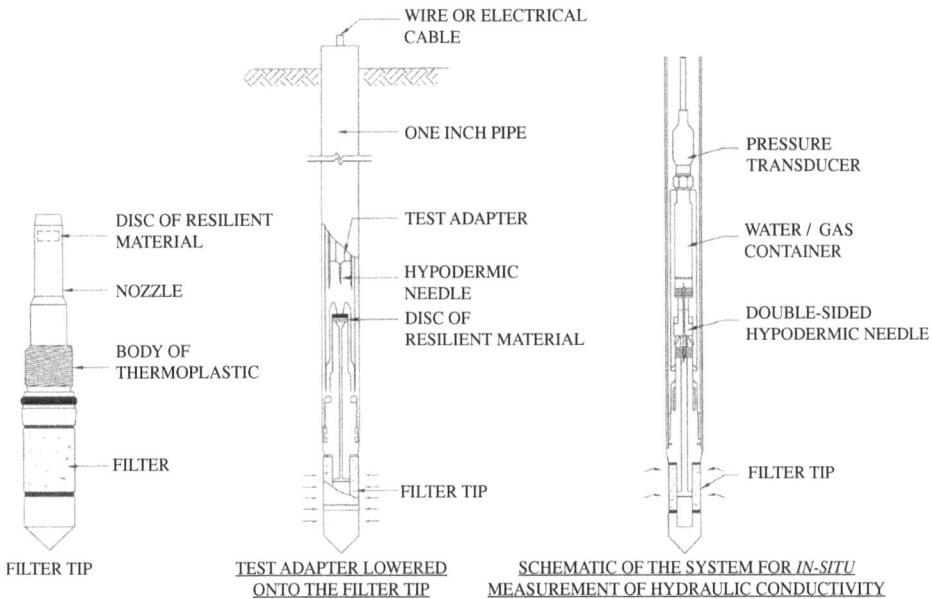

Figure 5.31   Geometry of a BAT permeameter.

Source: Bo *et al.* (2014).

$$k_h = \frac{P_0 V_0}{Ft}\left[\frac{1}{P_0 U_0} - \frac{1}{P_t U_0} - \frac{1}{U_0^2}\ln\left(\frac{P_0 - U_0}{P_0}\frac{P_t}{P_t - U_0}\right)\right] \qquad (5.21)$$

$$F = \frac{2\pi L}{\ln\left[\frac{L}{d} + \sqrt{1 + \left(\frac{L}{d}\right)^2}\right]} \qquad (5.22)$$

where k_h is the horizontal hydraulic conductivity in m/s; P_0 is the absolute initial system pressure; V_0 is the initial gas volume in ml; F is the shape factor and is calculated as 228.76 mm; U_0 is the static pore water pressure; P_t is the absolute pressure at time t; and L is the length of a filter in mm and d is the diameter of the filter in mm. All pressures are measured in meters.

Figure 5.32 shows a typical horizontal hydraulic conductivity versus elapsed time plot from the BAT permeameter test undertaken at a specific elevation. The example of horizontal hydraulic conductivity (k_h) values as obtained from the BAT permeameter tests is shown in Figure. 5.33.

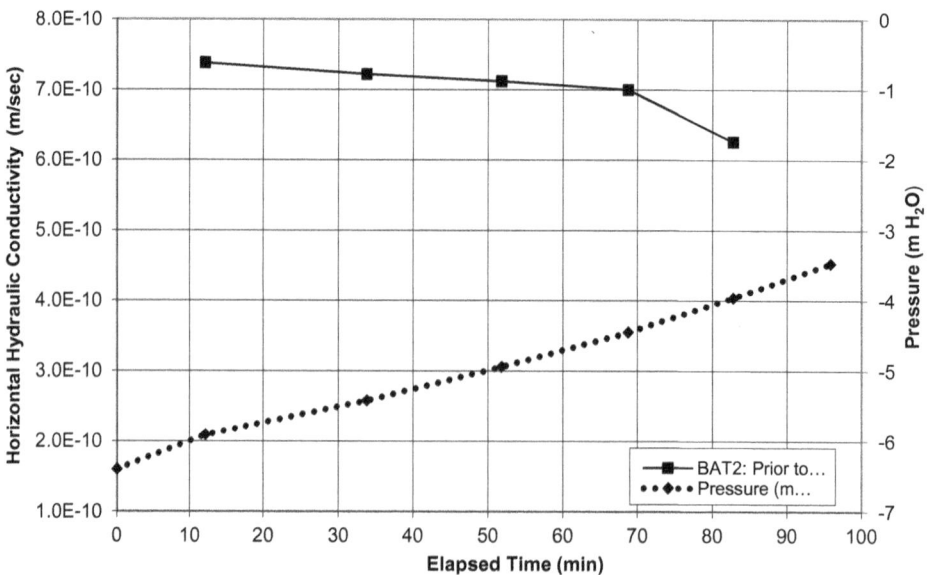

Figure 5.32   Typical horizontal hydraulic conductivity versus elapsed time plot from the BAT permeameter test.

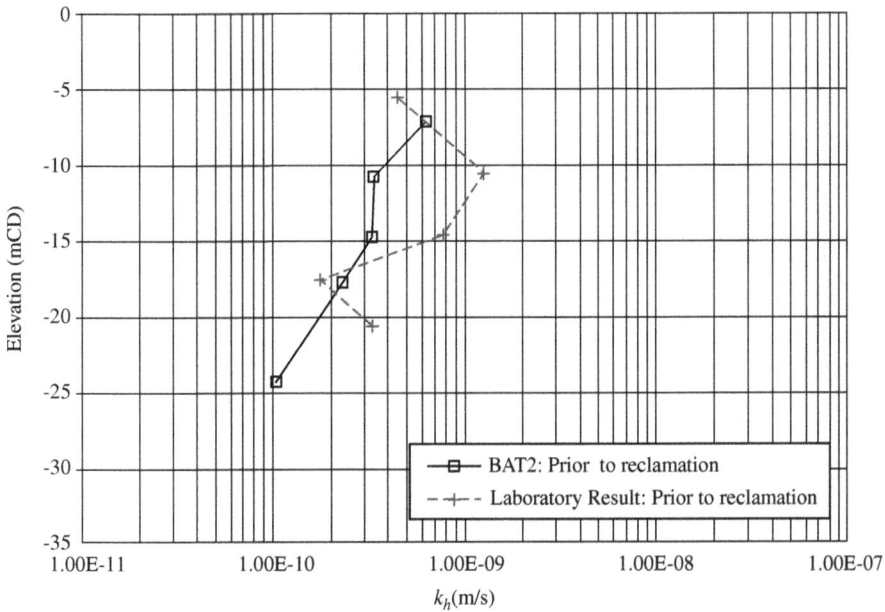

Figure 5.33   Horizontal hydraulic conductivity (k_h) values as obtained from the BAT permeameter tests.

5.9 Verification Testing

In addition to soil profiling and characterization of geotechnical parameters, geo-technical ground investigation methods are frequently used in the verification of achievement in ground improvement works. The most frequently used verification method for quality control of earthworks is compaction quality control. *In-situ* dry density and moisture content are usually measured *in-situ* to determine the achievement of the compaction process. The methods of measurement for dry density for compaction quality control are as follows:

- Core cutter method
- Sand Replacement Method
- Water displacement Method
- Balloon displacement method
- Nuclear Gauge

While nuclear gauge measures both dry density and moisture content *in situ* as described in Section 4.5, the core cutter method measures the volume using a core cutter mold which has a standard known volume and measures the weight of

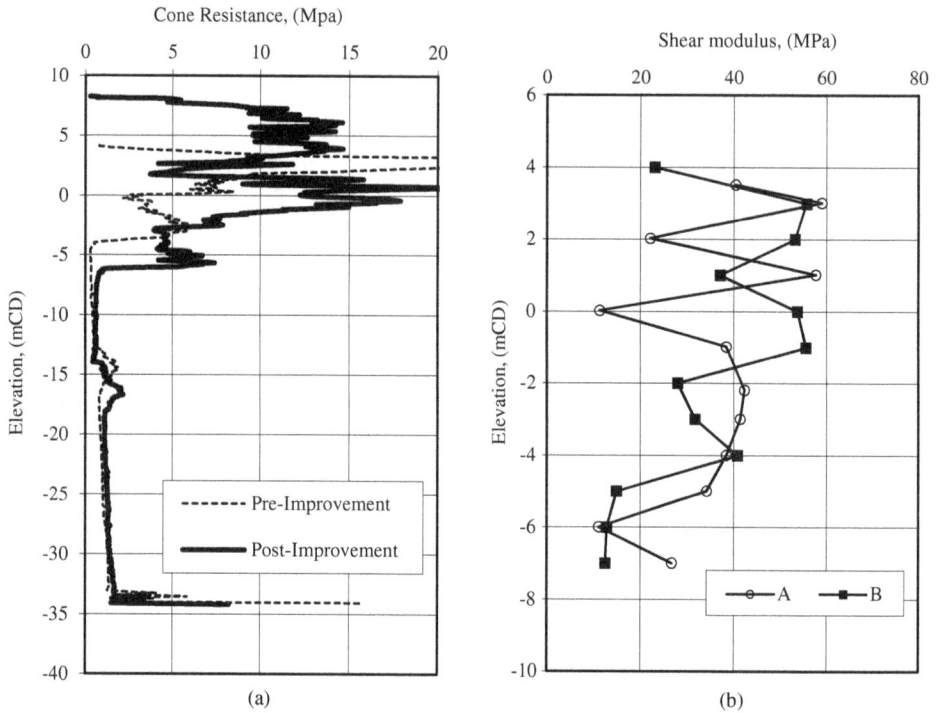

Figure 5.34   (a) Comparison of prior and post development cone resistance and (b) modulus using CPT (A) and CPMT (B) (right).

the soil *in-situ* or in the laboratory. Moisture content is also measured in the laboratory on the collected sample applying ASTM D 2216-19 and BS 1377-Part 1. Both water displacement and balloon methods measure the volume by replacing the excavated hole by filling the known volume of water or inflating the known volume of the balloon. Measured dry density and moisture content are verified against specified parameters. The sand replacement method uses the known volume of sand to measure the volume of soil.

The above-mentioned quality control tests are generally carried out over certain frequencies either per area (area project, e.g., land reclamation), per length (for linear project, e.g., Highway, Railway embankment), or per volume of fill (volume project, e.g., dam, landfill). Measurements are also carried out on every lift of the fill, which are generally less than 300 mm or limited to the influence depth produced by the compaction equipment used.

Improvement of granular soil can also be verified using CPTs (Bo *et al.* 2005) and CPMTs. Figure 5.34 shows a comparison of prior and post cone resistance measurements and modulus measurements using a CPT and CPMT.

Most deep compaction works are carried out over greater depth on the existing loose granular soils or land reclamation fill using the hydraulic filling method, and quality control using dry density tests at the various depths when the lift of the fill is gradually increasing is not possible. In such case, *in-situ* testing methods are usually applied to measure relative density. The most popular *in-situ* test for quality control is the CPT which can penetrate to the required depth by applying a static pushing force as described in Section 5.1. The CPT measures cone resistance and it is also possible to use it to interpret the classification of the type of soil penetrated. The achieved cone resistance against the specified cone resistance can be verified for the achievement of compaction works.

Alternatively, applications of the Standard Penetration Test and the dynamic cone test, called auto ram sounding test, are frequently used to verify quality control in deep compaction projects. The pressuremeter is frequently used to measure the improvement of modulus of elasticity of the soil.

In addition to the compaction and deep compaction of the ground, the underlying soft soils are frequently improved applying preloading with a prefabricated vertical drain to achieve the required degree of consolidation with specified future loads.

As degree of consolidation is directly related to effective stress gain or dissipation of excess pore pressure and increase in undrained shear strength, several methods of *in-situ* testing were used to verify the improvement (Bo and Choa 2004; Bo *et al.* 2019; Arulrajah *et al.* 2006a). Assessment of degree of consolidation and performance verification of soil improvement works using specialized *in-situ* test data were

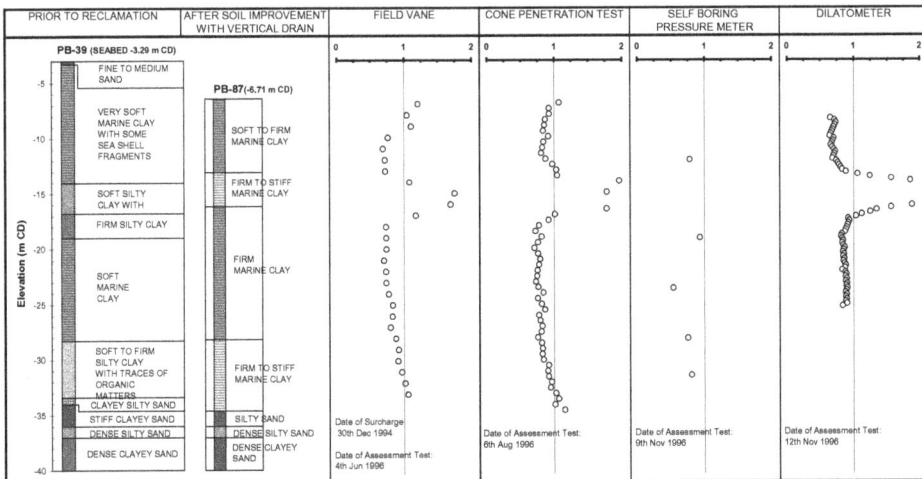

Figure 5.35   Degree of Consolidation measured by specialized *in-situ* testing.

Source: Bo *et al.* (1997).

extensively described by Bo and Choa (2000), Bo *et al.* (2012, 2014, 2015), and Arul *et al.* (2004a, 2008). Figure 5.35 shows the degree of consolidation measured after ground improvement work using specialized *in-situ* testing.

The following are other types of *in-situ* tests described in the previous section which are also used to verify the improvement of ground modification works:

- Field Vane Shear Test
- Cone Penetrometer with pore pressure measurements
- Dilatometer
- Self-Boring Pressuremeter
- Neutron Probe

All *in-situ* methods described above except the neutron probe test can either directly measure or interpret undrained shear strength and the overconsolidation ratio from measured parameters as described in the earlier Section. The status of soil improvement can then be monitored and verified. Figure 5.36 shows performance monitoring using *in-situ* testing in the Changi East Reclamation Project in Singapore in the late 1990s (Bo *et al.* 2003a).

In addition, change in soil conditions after ground improvement such as a change in hydraulic conductivity could also be detected by comparing prior and post measurements using specialized *in-situ* testing equipment.

As the neutron probe measures moisture content which is directly related to the void ratio for saturated soil, it is possible to verify the improvement of soil and achievement of the degree of consolidation from this measurement.

5.9.1 *Cone penetration test (CPT and CPTU)*

The cone penetration test can be carried out in the same manner as before the ground improvement. As explained in Section 5.1.1, the undrained shear strength can be estimated from the CPT. Figure 5.36 shows a comparison of undrained shear strengths measured by the CPT prior to and after improvement. Pore pressure measurements to determine the degree of pore pressure dissipation are used to estimate the OCR from the CPT.

Dissipation tests can be carried out in the same way as before the ground improvement. However, pore pressure should be normalized with equilibrium pore pressure obtained from the CPTU measurement rather than static pore pressure:

$$\text{Normalized pore pressure} = \frac{u_i - u_t}{u_i - u_e} \tag{5.23}$$

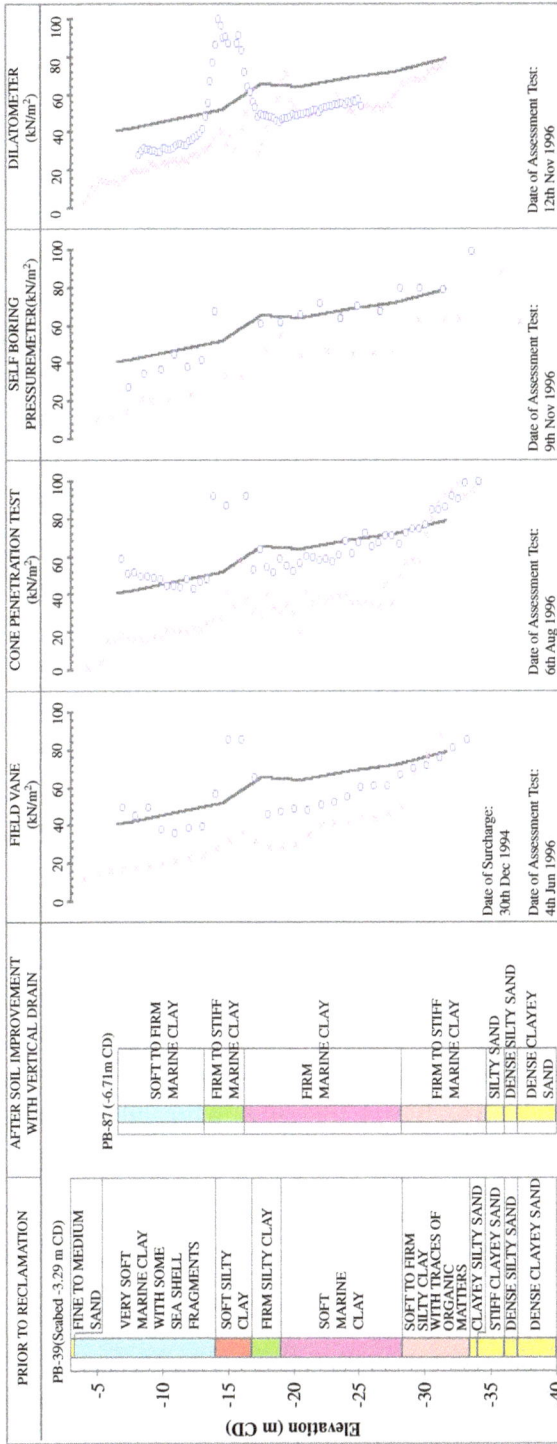

Figure 5.36 Performance monitoring data graphed using data obtained through *in-situ* tests.

Source: Bo *et al.* (1997).

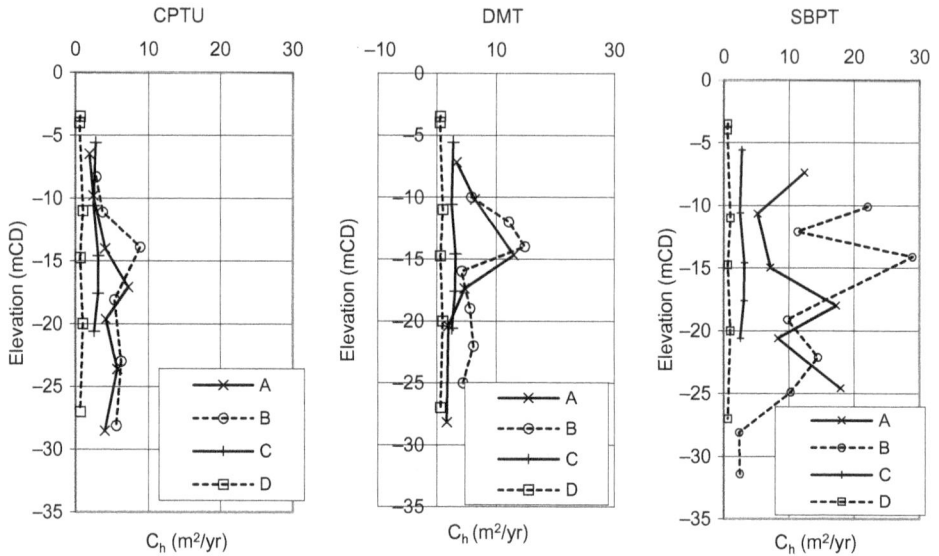

Figure 5.37   Comparison of C_h value measured before and after improvement from CPTU.

Note: A — Prior to reclamation *in-situ* results, B — After soil improvement results, C — Prior to reclamation laboratory results and D — Back analysis from field results.

where u_i is the initial pore pressure, u_t is the pore pressure at time t, and u_e is the equilibrium pore pressure measured from CPTU, DMT, and SBPT tests. Figure 5.37 shows a comparison of C_h values measured before and after improvement.

One more additional test can be carried out to check the improvement of the soil, which is long-term holding test. If the CPT cone is held at a certain elevation for a very long time, the pore pressure will dissipate to the equilibrium pore pressure. This equilibrium pore pressure will be the same as the pore pressure in the soil at the time of measurement. This will also be the same as the pore pressure measured with a piezometer. In that case, the degree of consolidation and effective stress can be estimated from a CPT long-term holding test. Figure 5.38 shows a comparison of equilibrium pore pressures measured using the CPT long-term holding test and that measured using a piezometer. It can be seen that the CPT long-term holding test measured the equilibrium pore pressure quite accurately.

5.9.2 *Dilatometer test*

A dilatometer test (DMT) can be repeated as post improvement *in-situ* testing. However, interpretation of most parameters from the DMT requires effective

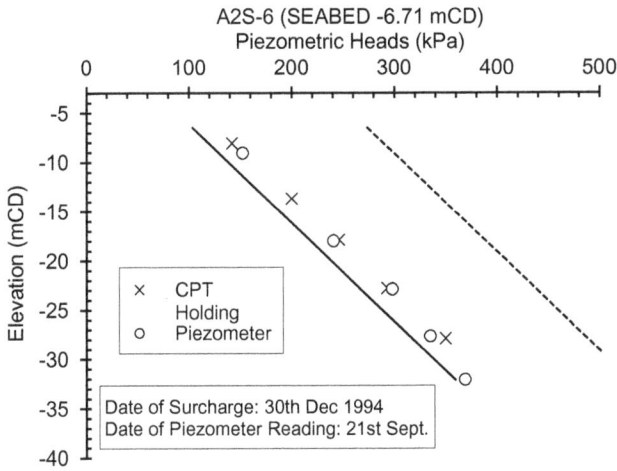

Figure 5.38 Equilibrium pore pressure from CPTU dissipation tests.

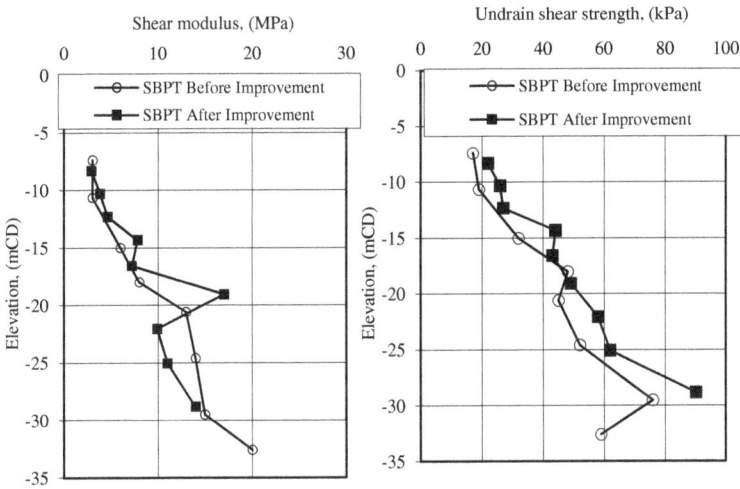

Figure 5.39 Comparison of soil parameters measured by the SBPT prior to reclamation and after ground improvement.

stress. As such, assessing improvement independently without knowing effective stress is impossible with a dilatometer test. Only an increase in modulus can be detected with a dilatometer since ED does not involve pore pressure and an effective stress term. The dilatometer dissipation test can also be reported to estimate the C_h values and K_h values after improvement. Figure 5.37 shows a comparison of C_h prior to and post improvement determined from DMTs.

5.9.3 *Self-boring pressuremeter test*

A self-boring pressuremeter test (SBPT) can be carried out in the same way as before the ground improvement, and the undrained shear strength and the OCR can also be estimated using Equations (5.16) and (5.17). The pore pressure dissipation test again can be carried out in the same way as described in Section 5.3.3. However, the equilibrium pore pressure should be used for normalization to obtain degree of dissipation as like in post improvement CPTU test.

Figure 5.39 shows a comparison of soil parameters measured by the SBPT prior to reclamation and after ground improvement. The dissipation test can again be carried out after improvement in order to interpret the C_h and K_h. Bo *et al.* (1997) have compared C_h and K_h from prior and post improvement using *in-situ* dissipation test data.

Chapter 6

Application of Geophysical Techniques in Geotechnical Ground Investigation

Geophysical investigation methods are extensively used in ground profiling during large-scale ground investigation works and also in site-specific profiling as well as *in-situ* testing for determining specific parameters. The following sections described surface geophysical techniques, borehole logging, and *in-situ* testing that are commonly used in geotechnical ground investigation.

6.1 Surface Geophysical Survey

The following surface geophysical investigation methods are frequently used in ground profiling and determination of groundwater levels for large-scale preliminary investigations (Zohdy *et al.* 1984):

- Seismic Reflection method
- Seismic Refraction method
- Resistivity survey

6.1.1 *Seismic surveying*

In seismic surveying, artificially generated pulses of energy are propagated through the soil mass and detected by seismometers or geophones at selected distances from the source of the pulse. These pulses are generally generated at or near the surface by detonating explosives or generating vibration using mechanical equipment. During the process, two parameters, such as velocity of these pulses and geometry of the propagation path, provide data to interpret several types of travelling waves between the source and the geophones. Each wave has a different transmission velocity and may also travel to geophones via more than one path. These paths include direct travelling along the surface and refracted or reflected paths from the boundaries between layers having two elastic properties or densities. The arrival of these paths may be recorded by different geophones at different distances.

These three types of ray paths may be summarized as follows:

(a) A direct surface ray travels horizontally along the surface from the source to the geophone.
(b) A completely reflected ray when a ray path strikes the boundary between two layers in which the deeper layer density is greater than the upper layer and when the angle of ray path from vertical (i) is greater than critical angle (i_c).
(c) A path of the ray striking the boundary precisely with a critical angle (i_c) and reflected back to the surface.

Using these ray paths and velocities, the depth to the boundary can be interpreted applying the following equation:

$$Z = \frac{1}{2}\sqrt{(V_1^2 * t^2 - X^2)} \qquad (6.1)$$

where Z is the depth to the boundary,
V_1 is the velocity along the upper formation,
t is the transmission time, and
X is the distance between the source and the geophone.

Details of the parameters are also shown in Figure 6.1. However, most seismic surveys are validated by confirmatory boreholes at selected locations. Details regarding seismic surveying and their interpretations can be found in Griffiths and King (1986).

In addition to profiling, the porosity and/or density of clastic sediments can be estimated from seismic surveys as there is a pronounced correlation between porosity and the density of sediments and compression seismic waves. While the velocity decreases with increasing porosity, velocity increases with increasing density. Eaton and Watkins (1970) developed a correlation between porosity and compression wave velocity.

6.1.2 *Resistivity surveying*

Resistivity Imaging Systems are generally used for shallow engineering and environmental applications. Specifically, the applications include the following:
(1) Engineering site investigations such as subsurface collapse features and conductive layer investigations for subway constructions, mines, and tunnel

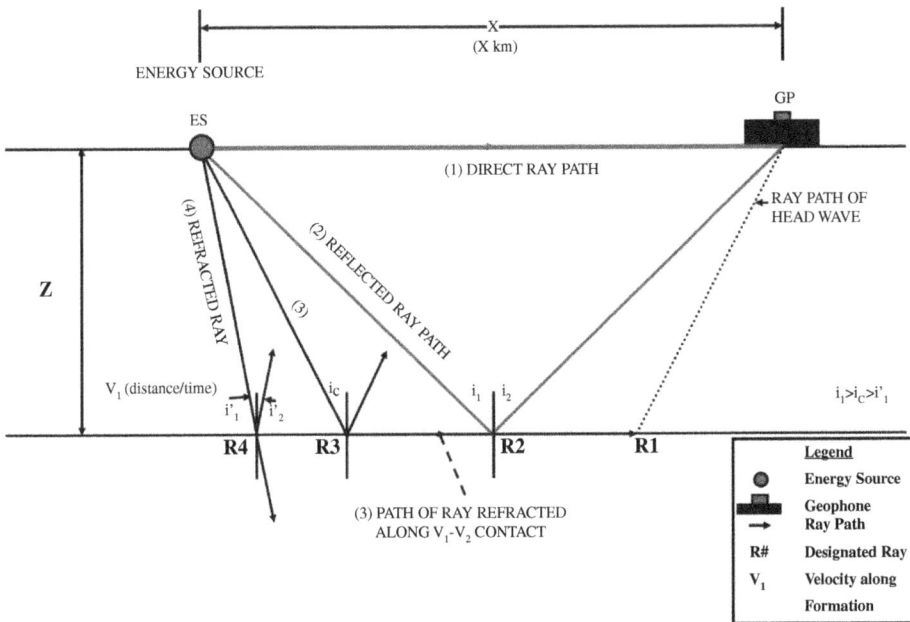

Figure 6.1   Application of seismic surveying using a seismometer/geophone and emitted ray path.

projects; (2) Environmental investigations such as closed landfills, potentially pol-
luted areas, leakage migrations, and contaminant plume mapping; (3) Groundwater
surveys such as detection of saline groundwater and groundwater potential; and
(4) Archaeological investigations for locations of buried objects.

In electric resistivity surveying, current is applied by conduction to the soil
through two electrodes and the potential differences between a second pair of
electrodes are measured. Electrical resistivity of the sub-surface layers is calculated from applied current and measured potential differences. From the variation
of electricity resistances, both vertical and horizontal variations of soil formation
can be interpreted.

Resistance measured between two potential surfaces separated by distance L
can be obtained from the following equation:

$$\rho = VA / IL \tag{6.2}$$

where ρ is the resistivity,
      V is the potential difference between two surfaces of constant potential,
      A is the cross-sectional area,
      I is the current in a conducting body, and
      L is the separated distance between two equipotential surfaces.

When current passes through the boundary of two media having differing resistivities, the lines of current flow will be refracted in accordance with the following equation:

$$\tan\theta_1/\tan\theta_2 = \rho_2/\rho_1 \tag{6.3}$$

where ρ_1 and ρ_2 are the resistivities of formation 1 and 2, respectively.
        θ_1 and θ_2 are the angles the current lines make with the normal and the boundary.

Resistivity measurements can be carried out using several configurations of electrode arrays. The following are configurations of arrays available in the industry, detailed descriptions of which can be found in Griffiths and King (1986):

- The four-electrode array
- Wenner configuration
- Schlumberger configuration
- Lee Partition configuration
- Dipole–dipole configuration
- Square configuration

The Wenner configuration which is frequently used in resistivity imaging is shown in Figure 6.2. Survey traverses are set out with electrodes placed at intervals of a ($a <= 5$m) as illustrated in the resistivity spread layout.

In practice, the system carries out sets of Wenner expansions along a traverse, using separations of a, $2a$, and na. The "n" value for a traverse changes with the traverse length, i.e., the longer a traverse, the wider an expansion. A current is transmitted into the ground through the outer two current electrodes of successive Wenner systems, and the two inner potential electrodes measure the potential difference between them. The diagram in Figure 6.3 also explains how measurements are made along a traverse and how the observed data are plotted as a conventional pseudosection at the points illustrated. The approximate survey depth would be half of the spacing of the potential electrodes, i.e., $na/2$.

In the installation of electrodes, steel electrodes are driven into the ground at designed spacings along survey traverses. If the surface to be surveyed is hard, a hand drill will be used to prepare holes for the electrodes. The steel electrodes are then connected to a multi-core cable which is in turn connected to the resistivity imaging system.

Before the start of a survey, contact between the electrodes and the ground is monitored. If the contact is not good, i.e., a high contact resistance is observed, a strong conductive brine solution would be used to moisten the ground surrounding the electrode and to reduce the contact resistance.

Switches inside the resistivity imaging system select groups of four electrodes at a time. An electric current is passed down the multi-core cable to the two outer electrodes in this group of four, allowing the current to flow through the ground

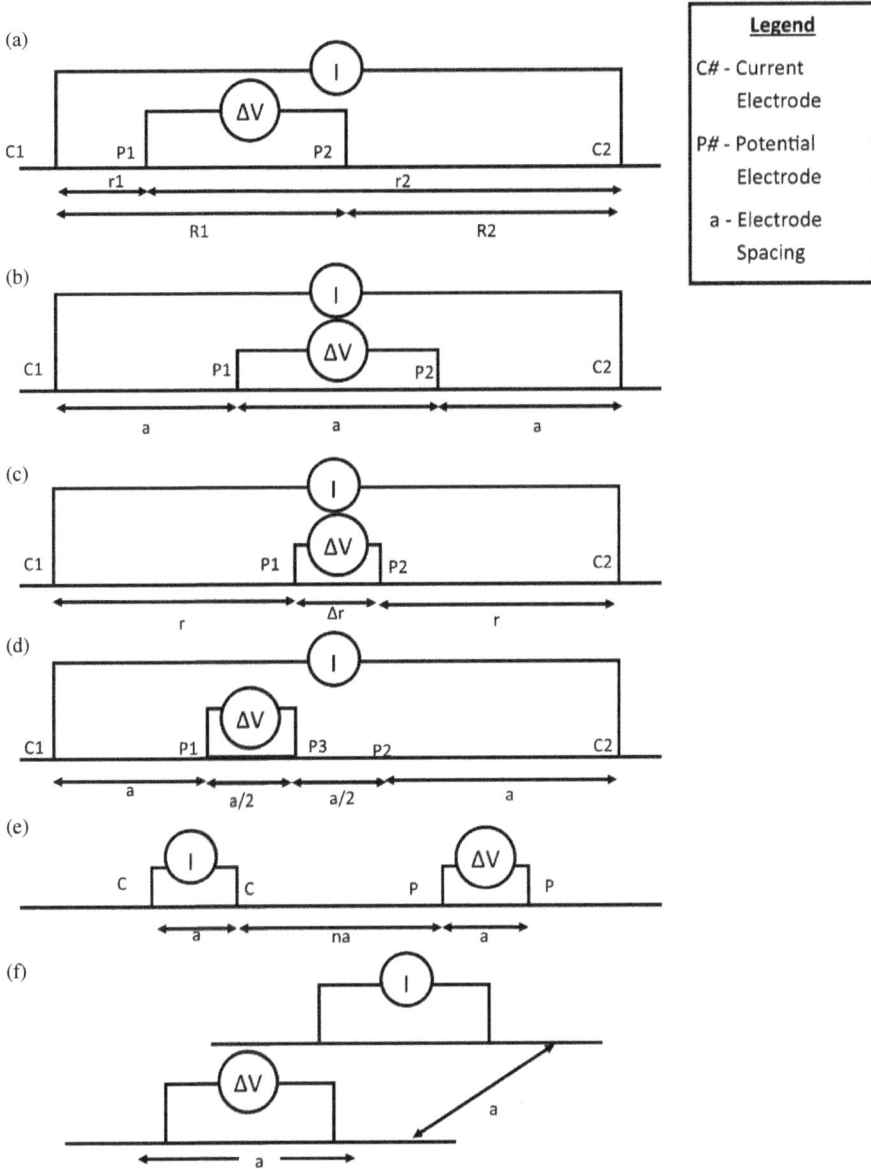

Figure 6.2   (a) The-four electrode array. (b) The Wenner configuration. (c) The Schlumberger configuration. (d) The Lee Partition configuration. (e) The Dipole–Dipole configuration. (f) The Square configuration.

Source: Griffiths and King (1986).

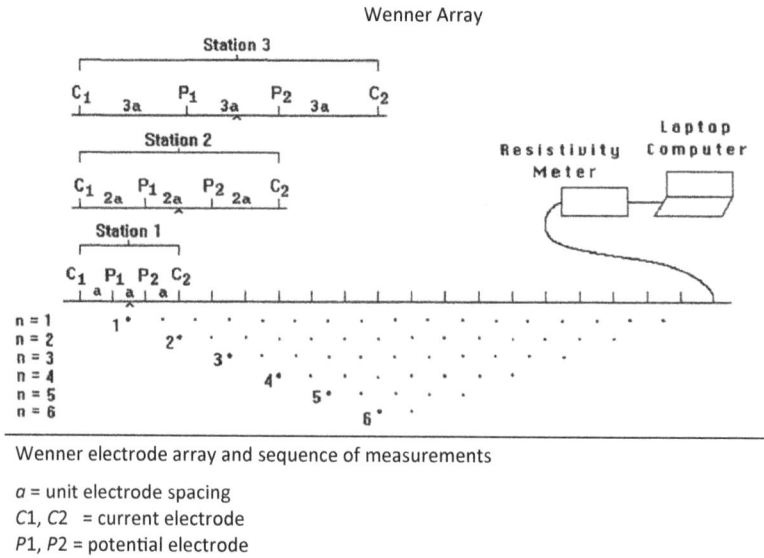

Figure 6.3   The Wenner electrode array and sequence of measurements.
Source: Courtesy of EGS.

between them. The potential difference is measured between the inner pair of electrodes, again through the multi-core cable.

The software driving the resistivity imaging system records the apparent resistivity of the ground from the current and potential difference measurements. The software then selects another group of four electrodes through the system switches and the process is repeated; this measurement process is repeated until measurements from all groups of electrodes at available separations are recorded.

Coordinates of electrodes are measured with standard topographic survey techniques if the surface to be surveyed is rough. Topographic correction will be carried out for the dataset.

Results of the resistivity survey are basically interpreted based on the resistivity distribution pattern of an inverse model and characteristics of rocks and soils at the survey site. Figure 6.4 shows the resistivity testing console and Figure 6.5 shows a resistivity survey in progress. Typical examples of resistivity survey results are presented in Figures 6.6 and 6.7 as cross-sectional and plan-view distributions, respectively.

6.2 Geophysical Logging

Geophysical logging within the borehole can be carried out in both open holes and cased holes depending upon the type of logging. Types of commonly used

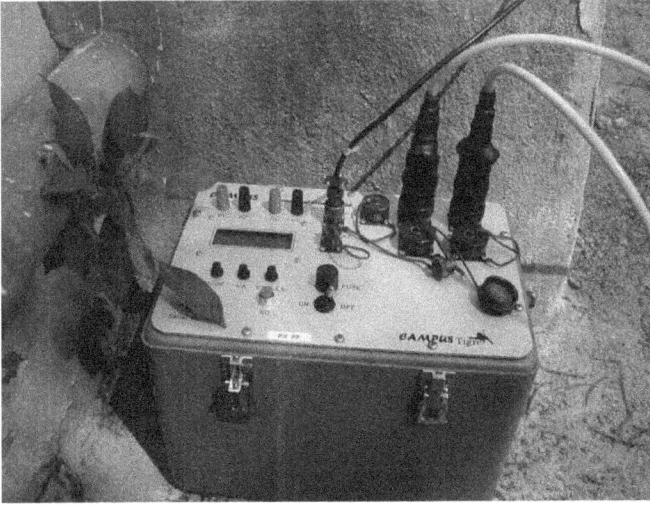

Figure 6.4   This photograph shows the Campus Tigre 64 console; a resistivity testing console.
Source: Courtesy of EGS.

Figure 6.5   Resistivity survey in progress, a survey carried out along the berm of a slope. Electrodes were attached to the multi-core (yellow).
Source: Courtesy of EGS.

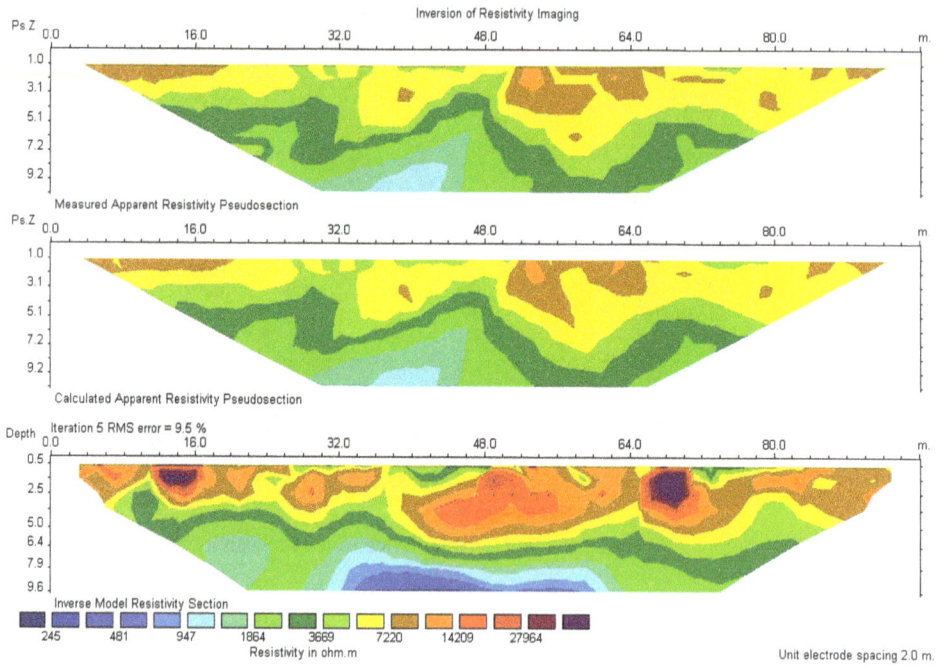

Figure 6.6   Resistivity inversion results. From top to bottom: Measured apparent resistivity, calculated apparent resistivity pseudosection, and an inverse model resistivity section.

Source: Courtesy of EGS.

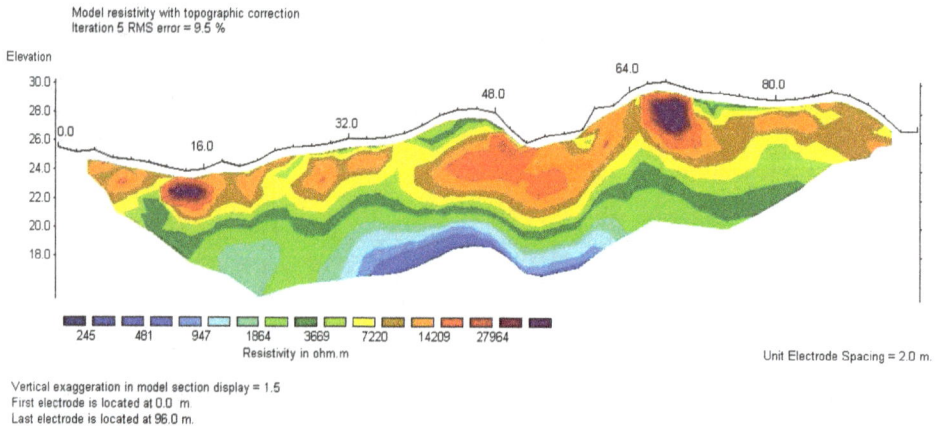

Figure 6.7   Model resistivity with topographic correction.

Source: Courtesy of EGS.

geophysical logging in the industry provide ground profiling and/or geotechnical characteristics of formations. They include the following:

- Resistivity probe
- Gamma–Gamma probe
- Neutron probe
- Spontaneous Potential Probe

These probes are available as individual probes or as multi-channels.

6.2.1 *Resistivity probe*

Resistivity probes measure the resistance of earth materials lying between two electrodes within the borehole. In reality, the measured potential differences between the two electrodes are converted to resistances applying Ohm's law using the following equation:

$$E = Ir \tag{6.4}$$

where E is the potential difference in volts,
I is the current in amperes, and
r is the resistance in ohms.

Resistivity probing can differentiate the type of soil and boundary conditions as different types of soils show different resistivities. In general, rock has high resistivity, whereas granular soil has higher resistivity than clayey soil. Normally, a qualitative logging methods are mostly used in the groundwater and petroleum industries. Figures 6.8(a) and (b) shows the resistivity electrode arrangement for core and borehole logging, and Figure 6.9 shows the current distribution for a differential resistance system.

6.2.2 *Gamma–Gamma probe*

Gamma–Gamma logs record the intensity of gamma radiation from the source, which is backscattered from the surrounding soils and rocks. Gamma–Gamma logs are utilized to classify the type of soils and rocks as well as the estimation of bulk density and porosity of soils and rocks. This probe consists of gamma protons, namely, cobalt-60 or cesium-137. This logging can be carried out either in an open hole or in a cased hole. The radius of the investigation is reported to be

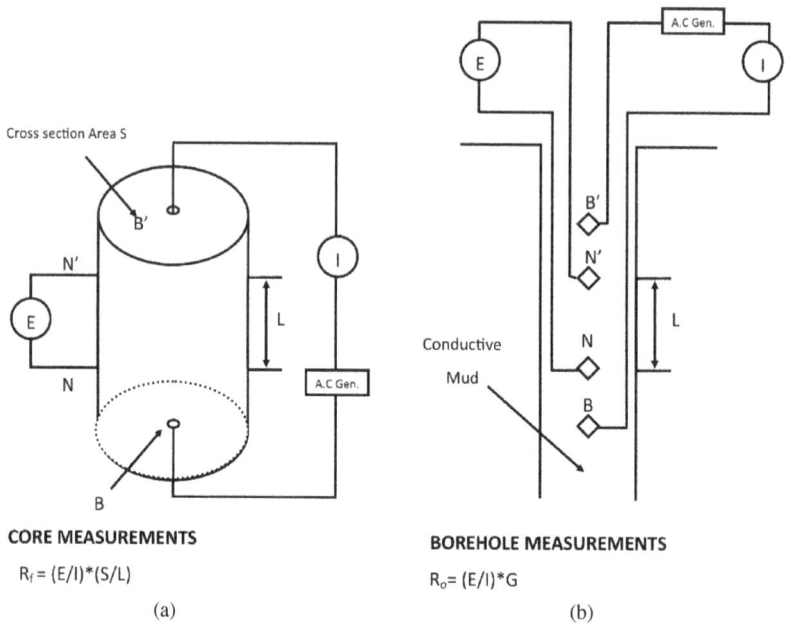

Figure 6.8   Electrode arrangement for resistivity measurements in (a) cores and in (b) boreholes. *Source*: Keys and MacCary (1985).

six inches around the hole. The count rate recorded from Gamma–Gamma logging is inversely proportionate to the bulk density as follows:

$$\phi = \rho_s - \rho_{bulk} \text{ (from log)}/\rho_s - \rho_w \tag{6.5}$$

Where ϕ is porosity, ρ_s is grain density, ρ_{bulk} is bulk density, ρ_w is fluid density.

Grain density is usually assumed to be 2.65 g/cc and fluid density is assumed to be 1 g/cc.

In practice, bulk density may be read directly from a calibrated and corrected log or derived from a chart providing the correction factor. Gamma–Gamma probes are usually calibrated in API limestone pits.

In addition to these quantitative measurements, the Gamma–Gamma probe is useful in identifying types of soils as some minerals are more responsive to radioactive gamma rays than others. For example, the clay mineral has more Gamma activity than quarts, which allows one to differentiate between clay and sand.

Figure 6.10 shows the equipment, principle, and interpretation of Gamma–Gamma logs.

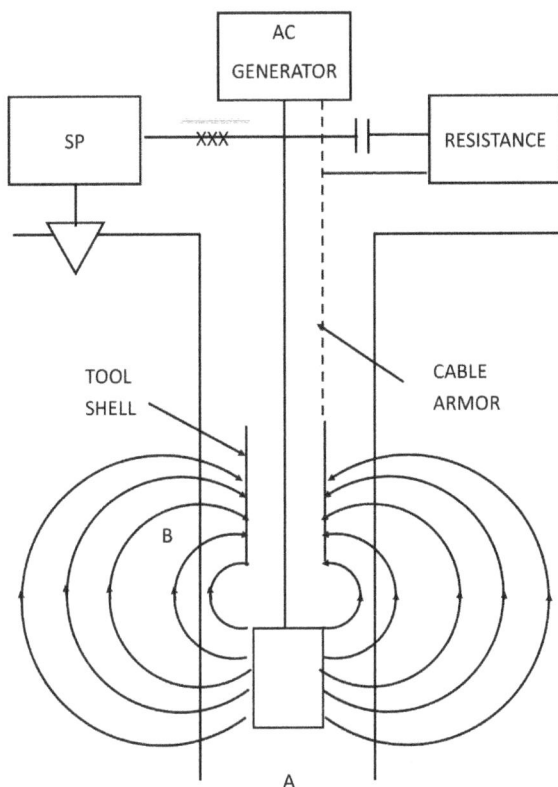

Figure 6.9   Differential resistance system, showing current distribution.
Source: Keys and MacCary (1985).

6.2.3 *Neutron probe*

Neutron probes use alpha particles impinging on beryllium. Neutron probes detect hydrogen content surrounding boreholes. Therefore, neutron probes can be used to measure the moisture content above the groundwater level and total porosity below the groundwater level as one of major contents in water is hydrogen. Neutron probes can be used in open holes, cased holes as well as dry holes or liquid-filled holes. The probe influences a relatively larger volume. Moisture probes and porosity probes use the same principle despite having some differences in the type of equipment and interpretation. Table 6.1 shows a comparison of two types of neutron probes. Figure 6.11 shows a typical arrangement of the Neutron probe and Figure 6.12 shows typical neutron and Gamma–Gamma probe logs and interpretation. The radius of investigation is reported to be between 6 inches and 24 inches depending upon the porosity of the soils and rocks. The lower the porosity, the higher the radius of influence.

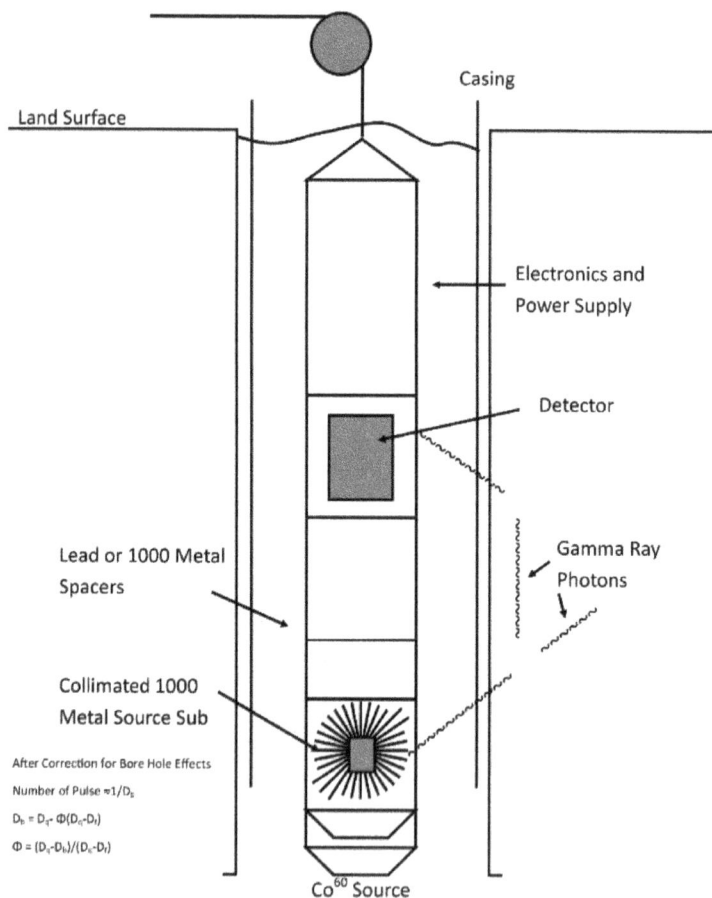

Figure 6.10   The equipment, principle, and interpretation of gamma–gamma logs.
Source: Keys and MacCary (1985).

6.2.4 *Spontaneous Potential probe*

Spontaneous Potential (SP) probes record the natural potential developed between the borehole fluid and the surrounding soils and rocks. A typical arrangement of the device is shown in Figure 6.13 (Lynch 1962). The device consists of a moveable lead electrode which can travel along the borehole with a probe and a ground lead electrode together with a device for measuring the potential in millivolts. This type of logging can determine formation thickness and differentiation between high and low permeable soils and rocks. This probe can only be used in the open hole which is filled with drilling fluid such as mud or water. The radius of investigation is highly variable and depends upon the conductivity of the intersected formation. The resulting data are plotted as the small differences in voltage in millivolts that developed between the drilling fluid and formation. In truth, the SP log is a measure of the potential drop that occurs in the drilling fluid. The SP is

Table 6.1   Comparison of neutron logging techniques.

	Neutron "Moisture Meter" Widely Used for Point Measurements of Moisture Content	Neutron "Porosity" Log Available on Oil Well and Some Water Well Loggers
Source-to-detector spacing	Usually less than 5 inches.	Usually more than 15 inches.
Source size	Millicurie range.	Millicurie range.
Type of detector	Usually sensitive to thermal neutrons.	Sensitive to thermal or epithermal neutrons or gamma rays.
Usual application	Moisture measurements.	Saturated porosity and moisture measurements.
Limitations	Usually restricted to small-diameter holes above the water table or air-filled pipes above the water table.	Useful in large- or small-diameter holes above or below the water table.
Advantages	Lower cost and reduced radiation exposure to personnel.	Wider application; reduced borehole effects; lower statistical error.
Graphic readout-neutron intensity	Increases to right.	Increases to right.
Moisture content	Linear response: moisture increases to right.	Logarithmic response: moisture increase to left.
Saturated porosity	Linear response: porosity increases to right; water in hole causes significant error.	Logarithmic response: porosity increases to left.

Source: Keys and MacCary (1985).

more positive in clayey soil and more negative in sandy soil as shown in a typical profile of a log (Figure 6.14).

6.2.5 *Sonic logging*

The P-S suspension logger is a low-frequency acoustic sonde designed to measure compressional and shear wave velocities (*Vp* and *Vs*) in soils and soft rock formations. It operates using indirect excitation rather than mode conversion as in a conventional sonic. It is capable of acquiring high-resolution *P* and *S* wave data in borehole depths of up to 600 m.

The *P–S* suspension sonde contains a unique design of a powerful hammer source and two receivers, separated by acoustic damping tubes. To acquire data, the sonde is stopped at the required depth and the source is fired under surface command. Firing causes a solenoid-operated shuttle aligned across the borehole axis to strike plates on opposite sides of the sonde in turn, setting up a pressure doublet in the surrounding fluid. The resultant fluid motion produces a tube wave at the borehole wall with

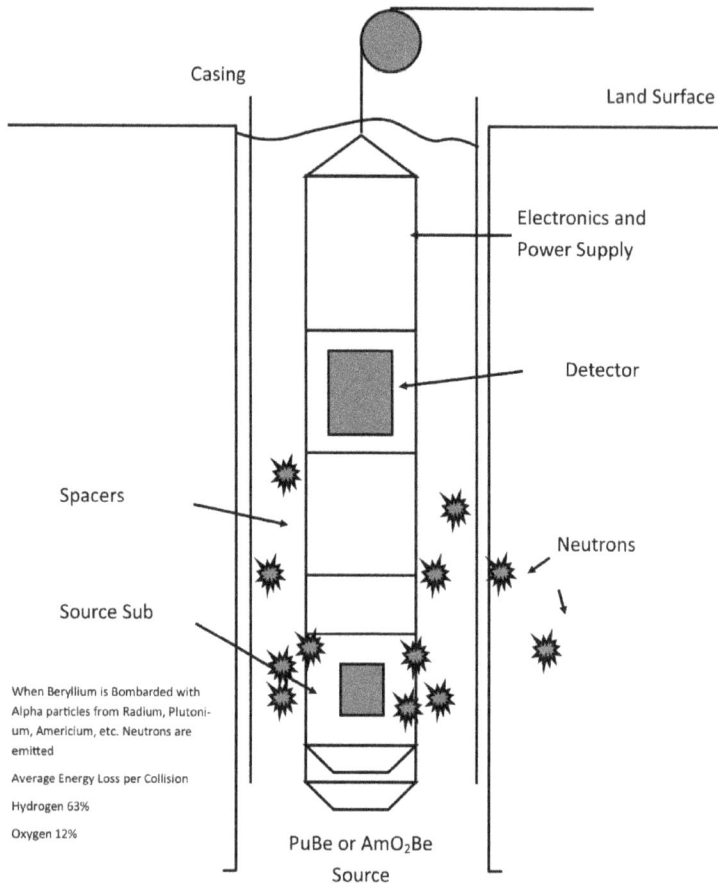

Casing

Land Surface

Electronics and
Power Supply

Detector

Spacers

Neutrons

Source Sub

When Beryllium is Bombarded with
Alpha particles from Radium, Plutoni-
um, Americium, etc. Neutrons are
emitted

Average Energy Loss per Collision

Hydrogen 63%

Oxygen 12%

PuBe or AmO$_2$Be
Source

Figure 6.11   The equipment, principle, and interpretation of neutron probing logs.
Source: Keys and MacCary (1985).

velocity close to the shear velocity of the formation together with a compressional wave. As the waves propagate parallel to the borehole axis, they set up corresponding fluid movements that are detected by the two neutral-buoyancy 3-D hydrophone receivers, allowing the wave velocity to be directly measured. The facility to stack multiple shots and filter the data as in normal seismic data acquisition is included in the operating software. Figure 6.15 shows details of sonic logging and Figure 6.16 shows the type of winch usually used in the pulling up and down of any logger probe, whereas Figure 6.17 shows geophysical logging in progress in the borehole.

The following seismic waves are recorded during logging:

- *P* wave data from both the near-receiver and far-receiver,
- A pair of *S* wave data from near-receiver (phase reversed),
- A pair of *S* wave data from far-receiver (phase reversed).

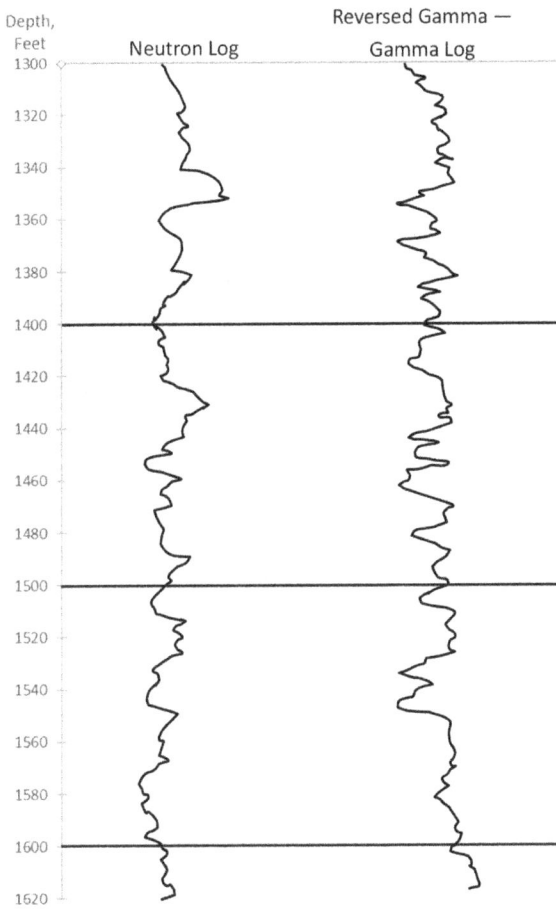

Figure 6.12   Typical neutron and Gamma–Gamma probe logs and interpretation.

Hence, for each trigger, there are three shots: 1 shot for P wave and 2 reversed shots for S waves. When necessary, the system can stack several shots of each polarity until the first arrival of the waves could be determined clearly. Travel times are then picked on site and approximate estimates of seismic velocities are made by the geophysicist to ensure the data quality. When a satisfactory record is obtained, the P–S logger is raised to the next level and measurements repeated. Measurements are carried out at vertical intervals, normally 1 m or 2 m.

The recorded data are analyzed, and the first break times of the P and S waves are picked, hence Vp and Vs. In some cases, post-acquisition filtering of the data can be used to make the first breaks easier to identify.

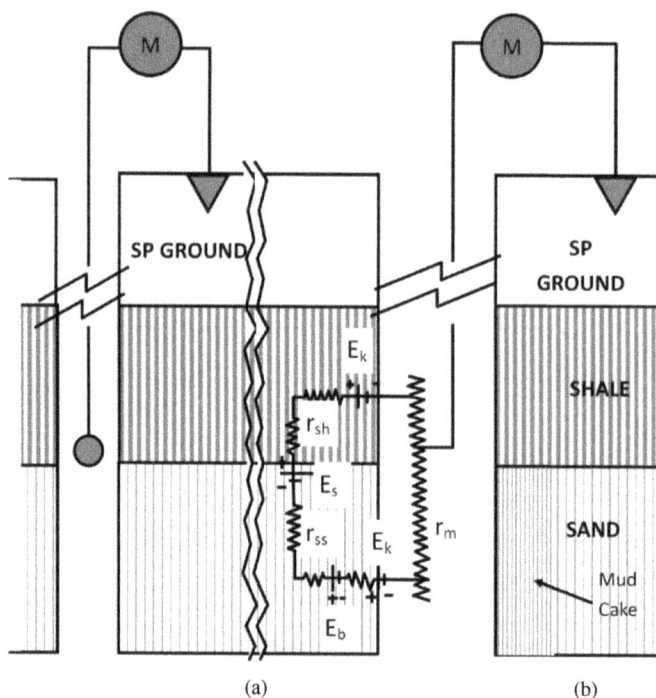

Figure 6.13   (a) Spontaneous potential measuring circuit and (b) electrical equivalent.
Source: Lynch (1962).

Given the bulk density, it is possible to calculate the following *in-situ* dynamic elastic moduli from the Vp and Vs measurements at each level:

E Young's modulus
σ Poisson's ratio
λ Lame's compressibility
k Bulk modulus
μ Shear modulus

6.3 Borehole Seismic

6.3.1 *Downhole seismic testing*

The downhole seismic test is an *in-situ* test and determines the velocity of primary (P) and secondary (S) seismic waves to provide the elastic soil parameters. The P and S waves velocities are used in geotechnical foundation analysis, static and

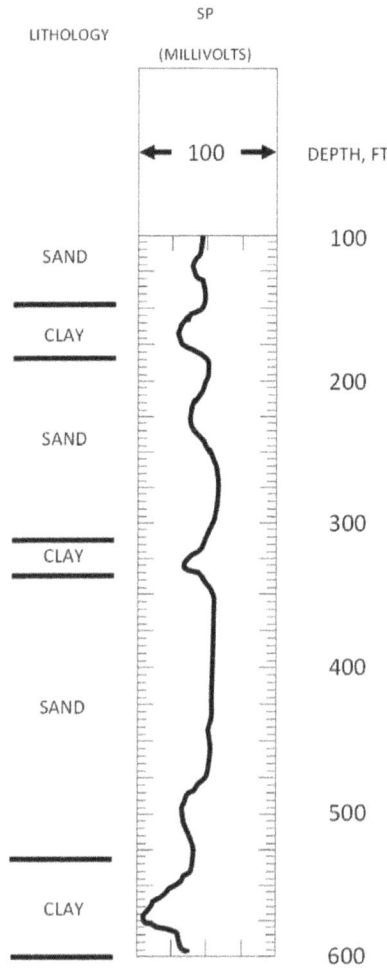

Figure 6.14 Typical SP log.

dynamic soil analysis, and liquefaction assessment. The downhole seismic test is carried out according to ASTM D7400. A wooden plank, a hammer, and a triaxial geophone will be required for this test. The borehole required for the testing is prepared according to the ASTM procedures using a PVD pipe. The wooden plank is placed on firm soil surface 3 m away from the borehole and soil bags are put on it for the stability. The geophone is connected with a computer by using a cable and placed at the bottom of the borehole. The S waves are generated by hitting separately at each end of the wooden plank with the hammer and the P waves are generated by hitting the wooden plank in a downward vertical direction for each test location. The velocities of both P and S waves are received by the geophones and recorded by using the specified computer program. The testing is carried out

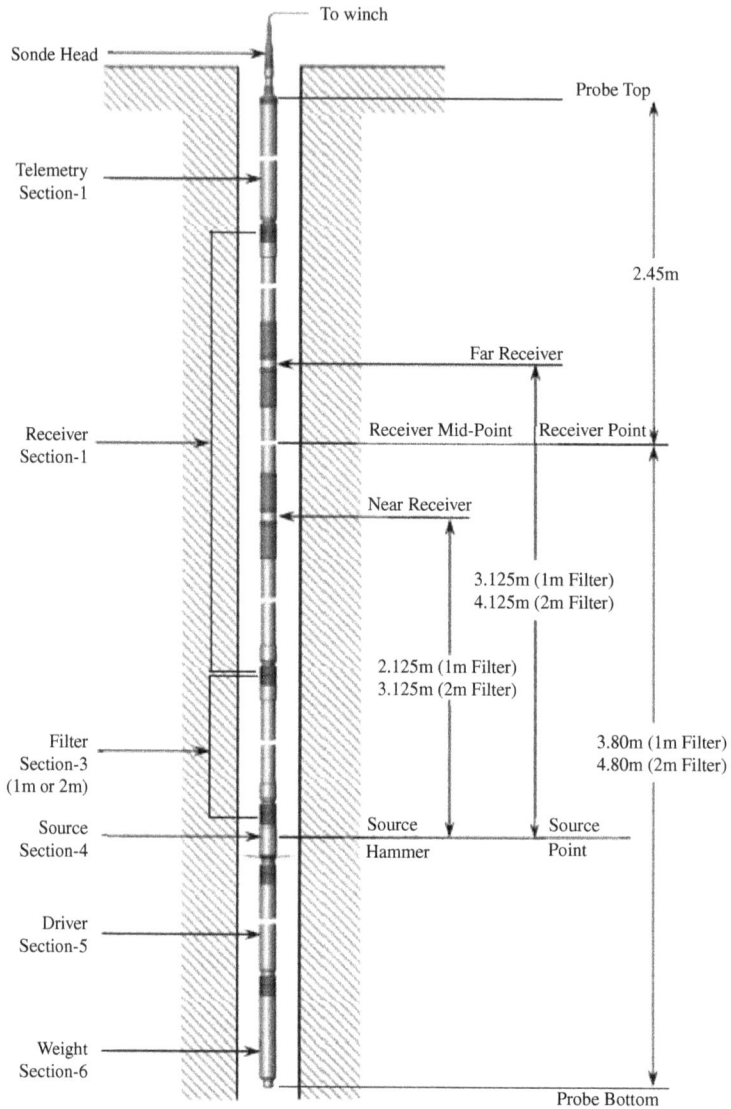

Figure 6.15   Sonic Logging apparatus and specifications.
Source: Courtesy of EGS.

up to the top of the borehole by lifting the geophone about 0.5 to 1 m above the previous test location with the same procedure. The depth of penetration can be up to 100 m. A time-depth graph and velocity-depth graph can be obtained for P and S waves. Based on the values of P and S waves, the Poisson's ratio and the constrained and shear modulus of the soil can be determined. Figure 6.18 shows a typical setup of downhole seismic testing.

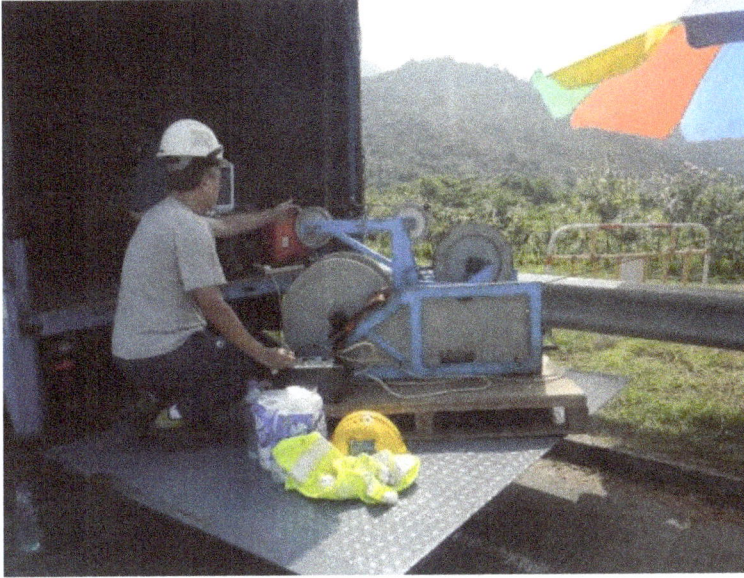

Figure 6.16   Typical winch that is used for pulling up and down of any logger probe.
Source: Courtesy of EGS.

Figure 6.17   Geophysical logging in progress in the borehole.
Source: Courtesy of EGS.

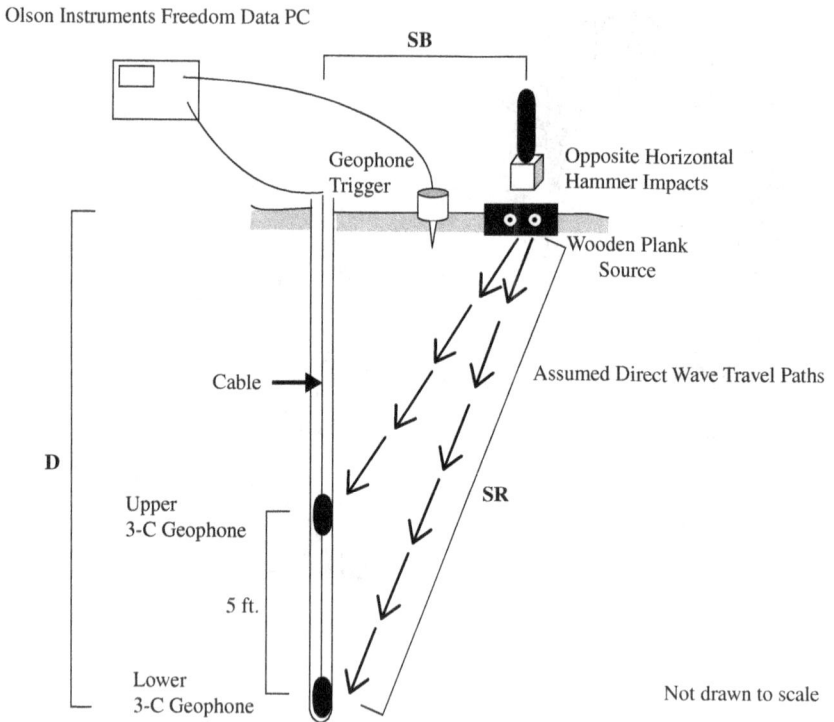

Figure 6.18   Downhole seismic test detail and typical setup.

6.3.2 *Crosshole seismic*

In crosshole seismic testing, the measurements of travel times for shear waves are made between two boreholes. In the testing process, a shear wave hammer (sound source) would be lowered down to the bottom of one of the boreholes (the "shot hole") and fired in both the up and down directions. The shear wave hammer has a contact closure system which will trigger the seismograph. This is an instantaneous trigger with zero delay. The hammer has an additional advantage. It can generate a shear wave in both the up and down directions; this will change the polarity of the shear wave and allow the shear wave to be easily identified on the seismogram. The seismic signals would be received by a three-component geophone at the same level in another borehole (the "receiver hole") and recorded on a seismograph. Figure 6.19 shows the arrangement of cross-hole seismic measurements and Figure 6.20 shows an example of seismic records.

The sound source and receiver would then be raised by a selected increment within the borehole (such as 1 m or 2 m, to suit the project requirements) and the

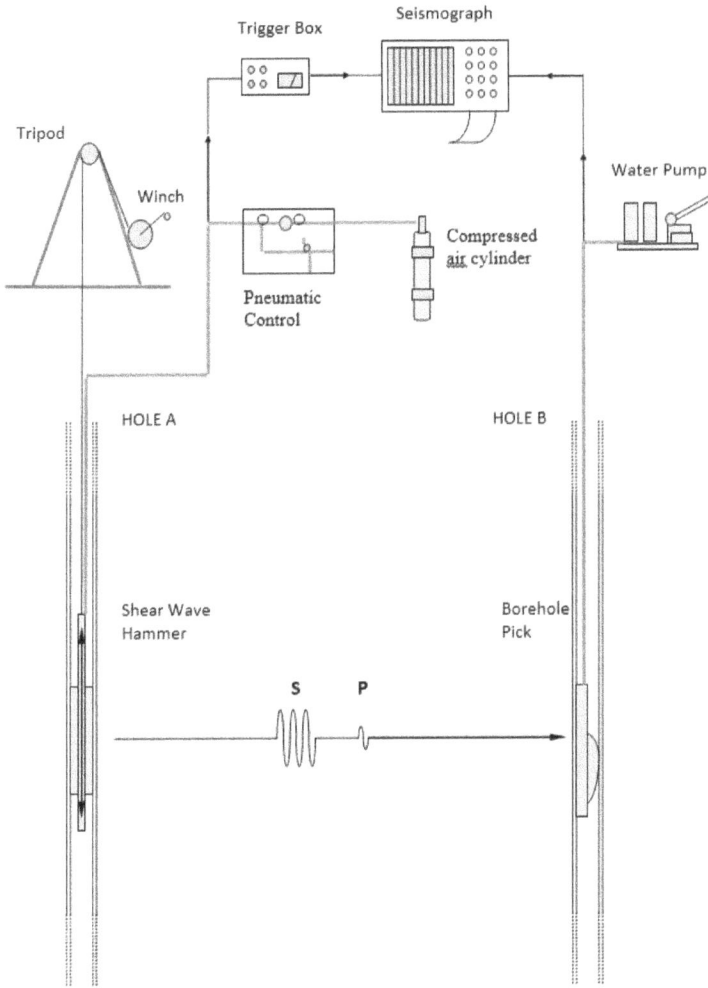

Figure 6.19   Arrangement of crosshole seismic measurements.
Source: Courtesy of EGS.

measurements repeated. Measurements will be recorded from the base of the zone of interest to the top of the zone of interest. The travel times and borehole separation would be used to calculate the compressional and shear wave velocities at each measurement depth.

The small-strain limit dynamic elastic moduli can be calculated using seismic wave measurements. The calculation requires the bulk density of the soil/rock around the boreholes. For the most accurate results, density should be measured in

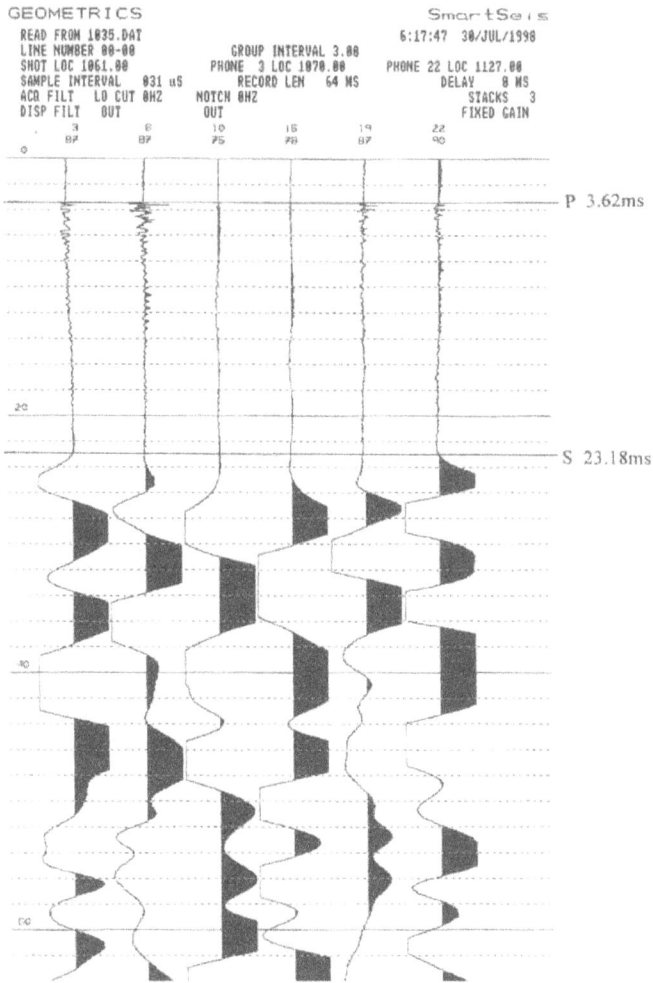

Figure 6.20   An example of a seismic record is shown.
Source: Courtesy of EGS.

a geotechnical laboratory from undisturbed samples collected during drilling. It is less accurate, but if measured density is not available, the density can be estimated based on the type of soils and rocks from available samples and the borehole log.

Given the bulk density, it is possible to calculate the following *in-situ* dynamic elastic moduli from the Vp and Vs measurements at each level:

E	Young's modulus
σ (sigma)	Poisson's ratio
λ (lambda)	Lame's compressibility

| k | Bulk modulus |
| μ (mu) | Shear modulus |

6.4 Multi-channel Analysis of Surface Wave

When seismic waves are generated at or near the surface of the earth, both body (P and S) and surface (e.g., Rayleigh and Love) waves are generated (Park *et al.* 1999). While body waves propagate through the whole body of the earth, the surface waves propagate along the surface of the earth. Surface waves have dispersion characteristics, while body waves do not have such characteristics. These dispersion characteristics make the surface waves with different wavelengths penetrate different depths with different velocities. Theoretically, short wavelengths penetrate shallow depths and longer ones penetrate deeper depths. Therefore, analyzing the dispersion of surface (shear) waves allows one to obtain a near-surface velocity profile (Park *et al.* 1997). These measurements can be implemented by creating seismic forces at the source and measuring shear wave velocities from the surface using several arrays of geophones with equal spacing through a multi-channel acquisition unit. After acquiring the data, the process of the analysis involves acquisition of dispersive data, construction of a dispersion curve, and back calculation of Shear wave velocities from the constructed dispersion curve. Various types of multi-channel data processing techniques developed by many researchers can be used (Herrmann 1973; McMechan and Yedlin 1981; Park 1997).

A method developed by Park *et al.* (1997) called Multi-Channel Analysis of Surface Waves (MASW) is usually applied. Theoretically, both types of surface waves such as Rayleigh and Love waves can be calculated by using following elastic wave equation (Haskell 1953):

$$\frac{\partial^2 \Phi}{\partial t^2} = v_p^2 \nabla^2 \Phi \tag{6.5}$$

$$\frac{\partial^2 \Psi}{\partial t^2} = v_s^2 \nabla^2 \Psi \tag{6.6}$$

where Φ and ψ are displacement potentials and v_p and v_s are velocities of P and S wave, respectively.

Shear moduli (μ) can be calculated using the following equation:

$$V_s = \sqrt{\frac{\mu}{\rho}} \ or \ \mu = v_s^2 \rho \tag{6.7}$$

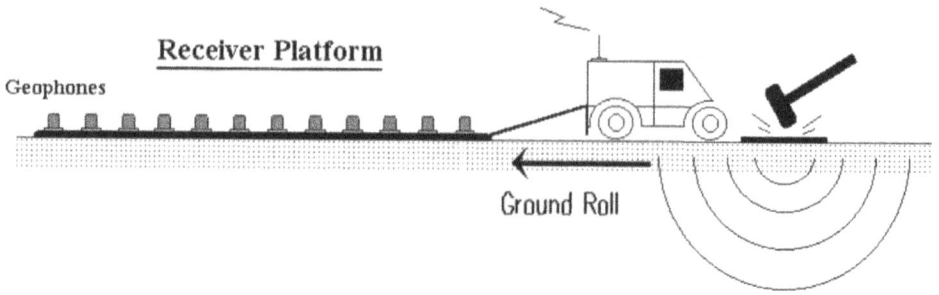

Figure 6.21   MASW Operating Principle.
Source: Courtesy of Park (1997).

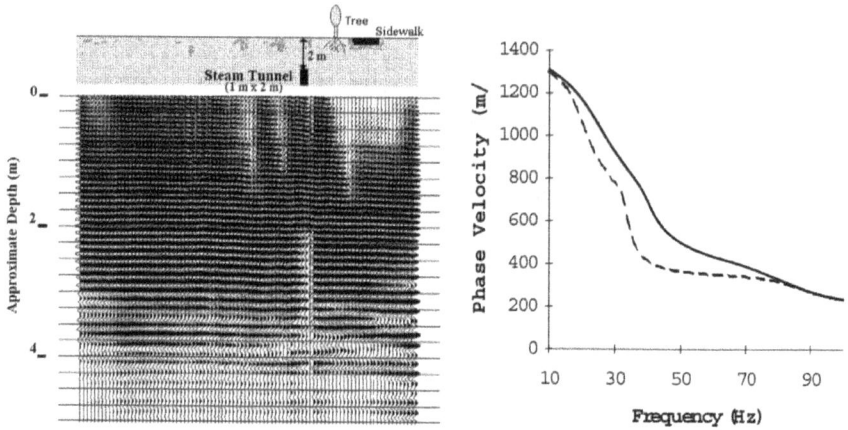

Figure 6.22   Example of a typical MASW shot record and phase velocity/frequency.
Source: Courtesy Park (1997).

Normal Zone

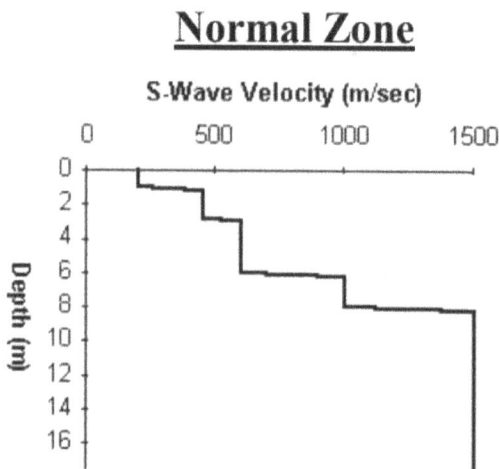

Figure 6.23   MASW shear wave velocity sounding.
Source: Courtesy of Park (1997).

where ρ is the density of materials and u is the shear modulus.

This measurement process can be carried out without necessarily advancing the borehole. The investigation usually applies the refraction methods to generate a shear wave velocity model as shown in Figure 6.21.

The shear wave velocity profile obtained from this analysis is useful in geotechnical and foundation engineering as the shear wave velocity is directly proportional to shear modulus, which indicates stiffness of the material.

The dispersion properties are measured as a change in phase velocity with frequency. The surface wave energy will decay exponentially with depth. Figure 6.21 outlines the basic operating procedure for the MASW method. Figure 6.22 shows an example of a typical MASW record and the resulting 1-D *Vs* model. Figure 6.23 shows measured shear wave velocity vs depth. A more detailed description of the method can be found in Park *et al.* (1997).

Chapter 7

Sampling

When geotechnical investigations are carried out for engineering purposes, collections of suitable soil, rock, and groundwater samples using appropriate methods are required. These soil and rock samples are normally brought back to the geotechnical laboratory in house or sent to other accredited third-party geotechnical laboratories for further testing. Groundwater samples are sent to an accredited water chemistry laboratory. These soil and rock samples may also be used for visualizing the confirmation and updating of the classifications made in the field during investigations. This chapter will describe the entire process of planning of sampling to collection of suitable types of samples, and the process of packaging, identifications, transportation, storage, and preparation for testing to obtain necessary geotechnical parameters. This chapter will also describe types of samples, types of available samplers in the market, and causes of sample disturbances.

7.1 Planning for Sampling

Before leaving for the geotechnical investigation sites, the field supervising geotechnical engineer should prepare types of samplers required depending upon the type of soils and rocks which are being investigating and the method of investigation, such as test pitting, soil drilling, and rock coring. For packing of the disturbed samples and groundwater sampling, suitable containers, sealable plastic bags, and jars may be required. Undisturbed samples may require suitable sealing materials such as wax, and for the rock core samples, prefabricated core boxes may be required. Sampling intervals depending upon types of soils and rocks, depths of investigation holes, and expected formation changes should be planned, and one should bring sufficient numbers of samplers, packaging materials, etc.

7.2 Types of Sampling

There are two types of soil samples, rock core samples, and groundwater samples which are required to be collected during ground investigation.

7.2.1 *Disturbed sample*

A disturbed sample can be obtained from test pitting, trenching, or from boreholes drilling. Disturbed samples can be obtained from sludge returns from direct or reverse circulation rotary drilling during advancing of boreholes, whereas returns from auger drilling could be collected as disturbed samples when using either hollow or solid stem augers. In percussive drilling, disturbed samples can be obtained using the bailing of cuttings resulting from the drilling. In the United Kingdom, window samplers which can be either pushed in with static forces or driven in with dynamic forces are frequently used to obtain disturbed samples. Window samplers are frequently used to obtain a continuous profiling and sampling of disturbed samples. From the test pits and trenches, disturbed samples can be obtained from a scoop of the excavator or collected using a shovel during excavation at the depths of interest. SPT samplers called split spoon samplers are frequently used to obtain disturbed samples for both cohesive soils and granular soils during the implementation of SPT testing. While good recovery of SPT samples can be obtained from firm to stiff cohesive soils and cemented granular soils, recovery in collecting soft to very soft cohesive soils and loose to very loose granular soils is limited with the SPT sampler. The method of SPT sampling is described in detail in ASTM D 1586. The disturbed samples are usually completely remolded, and they may also not be in their states of natural strength and natural water content. Disturbed samples are only useful for some basics classification tests such as Atterberg limit tests and grain size distribution tests. They may not even be suitable for natural moisture content tests. When tests such as natural moisture content tests, unit weight, strength, and consolidation tests are required, undisturbed samples (intact samples) will be required to be collected. Table 7.1 shows different classes of samples and types of tests which can be carried out from particular types of samples, with Class 1 being the best-quality sample which is required for strength and deformation tests and Class 5 being the lowest-quality sample which is only suitable for visual inspection and not suitable for any kind of laboratory test. Table 7.2 also shows the quantity of soil samples required for various types of classification tests and compaction tests.

7.2.2 *Undisturbed sample*

To obtain undisturbed samples, soil should have a suitable degree of cohesion or cementation; otherwise, the possibility of obtaining undisturbed sample is poor

Table 7.1   Sample quality: BS 5930 1981.

	Classification	Moisture Content	Density	Strength	Deformation & Consolidation
Class 1	•	•	•	•	•
Class 2	•	•	•		
Class 3	•	•			
Class 4	•				
Class 5					

Table 7.2   Mass of soil sample required for various laboratory tests BS 5930 1981.

Purpose of Sample	Soil	Mass of Sample Required (Kg)
Soil identification, including Atterberg limits, sieve analysis, moisture content, and sulfate content tests.	Clay, silt, sand Fine and medium gravel Coarse gravel	1 5 30
Compaction tests.	All	25–60
Comprehensive examination of construction materials, including soil stabilization.	Clay, silt, sand Fine and medium gravel Coarse gravel	100 130 160

or slim. Careful methods of sampling using a suitable tool which can prevent disturbances during sampling or minimize the disturbances are deemed necessary. This mean undisturbed sampling is possible for cohesive clayey soils with a certain magnitude of *in-situ* strength. Despite some cohesion being present, if the samples do not have sufficient *in-situ* strength, they could be easily disturbed. Very good undisturbed block samples can be obtained from the bottom of excavated test pits and/or trenches by hand excavated sampling. Figure 7.1 shows block sampling during test pitting. Block samples can only be taken from depths in heavily overconsolidated soils, such as London clay. In normally and lightly overconsolidated clays, excavation of a pit or shaft to more than a few meters depth is often impossible because base heave will occur. Lefebvre and Poulin (1979) calculate that, for example, in clay with undrained shear strength, the depth of a trench or pit will be limited to about 4 m, if a factor of safety of two is to be maintained.

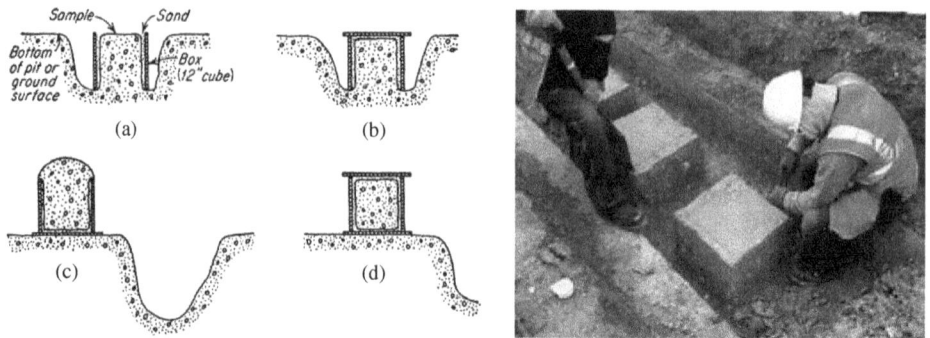

Figure 7.1   Block sampling in cohesive soil.

Undisturbed samples can also be collected during drilling operation. Table 7.3 shows classes of sample quality which are obtainable for the respective types of soil using specific methods of drilling investigation and using suitable samplers. It can be seen that only a U100 sampler and piston sampler can obtain a Class 1 sample which is required for strength, deformation, and consolidation tests. Normally, consolidated and lightly consolidated samples required a piston sampler to obtain a Class 1 sample. Overconsolidated clay can be sampled with a U100 sampler to obtain a Class 1 sample due to its greater strength than normally consolidated clay. If clay contains gravels, cobbles, and boulders, it will prevent one from obtaining a Class 1 sample even with a piston sampler. Non-cohesive granular soil such as sand and silt cannot obtain a Class 1 quality sample even using a piston sampler. Most methods require undisturbed sampling using a U100 or piston sampler after retrieving the entire drill string each time sampling is required. Drilling using a hollow stem auger allows sampling through the hollow stem drill rod without the necessity to retrieve the entire drill string each time sampling is required.

In addition to the U100 and piston sampler, there are a few types of large-diameter samplers available in the market which could provide better-quality samples and these will be described in the subsequent section.

To obtain a rock sample, a collection of core samples using a core barrel applying the rotary drilling method is required. There are core barrels with single-, double-, and triple-tube liners inside the barrel. When better-quality cores are required, double- and triple-tube core barrels are used. Friable and water-resistant soft rock like shale requires a triple-tube core barrel to obtain better-quality cores with minimum sample disturbances. Wireline coring allows continuous core sampling without retrieving the entire string of drill assembly by sampling through the hollow drill rods.

Table 7.3   Quality of samples obtainable, according to soil type information from BS5930: 1981.

Soil Type	Method	Sample Quality Class
Non-cohesive soils containing boulders, cobbles, or gravel	Disturbed samples from dry trial pit	4
	Disturbed samples from shell during light percussion drilling	5
Sand	Disturbed samples from a shell during light percussion drilling	5
	SPT split spoon	4
	U100 with core catcher	4
	Piston sampler (medium dense sand)	2
	Piston sampler (loose to very dense sand)	3
	Compressed air sand sampler	As piston sampler
Silt	Shell or clay cutter	5
	U100	3
	Thin-walled piston sampler, in low- or medium-density silt	2
Normally consolidated and lightly Overconsolidated clays	Clay cutter	4
	Shell (intact lumps)	4
	Open-tube samplers	2
	Thin-walled piston sampler	1
Overconsolidated clays	U100 — for firm to stiff clays —	1
	very stiff or hard, particularly if they are closely fissured or stony	2 or 3
Clay containing gravel, cobbles, or boulders	Disturbed samples from dry trial pit	3
	Shell	5
	U100	4 or 5
	Rotary core	4
Rock	Shell	5
	Clay Cutter	4 or 5
	U100	3 or 4
	Rotary Core	Not given

7.2.3 *Rock sampling*

Rock sampling is usually carried out during rotary drilling which was described in Chapter 4. Rock cutting and penetrating into the rock masses are carried out using a diamond bit attached to the core barrel which is the sampler used for retrieving the rock core. There are several types of core barrels such as single, double, and triple-tube core barrels. While the double-tube core barrel provides better-quality core samples, the triple-tube core barrel provides the best-quality core samples. For weak, friable, and water-sensitive rock, sampling requires a minimum of a double-tube core barrel but preferably a triple-tube core barrel. Figure 7.2 shows various types of single-, double-, and triple-tube core barrels. There are many different sizes of core barrels which are available in the market are shown in Table 7.4. The most commonly used sizes of core barrels in the geotechnical industry are NQ and HQ core barrels. The methods of core sampling are extensively described in BS5930 (1999) and ASTM D 2113-02 (2006).

7.2.4 *Groundwater sampling*

Many times, groundwater sampling is required to carry out the groundwater chemistry tests to assess the aggressiveness of groundwater which could attack the

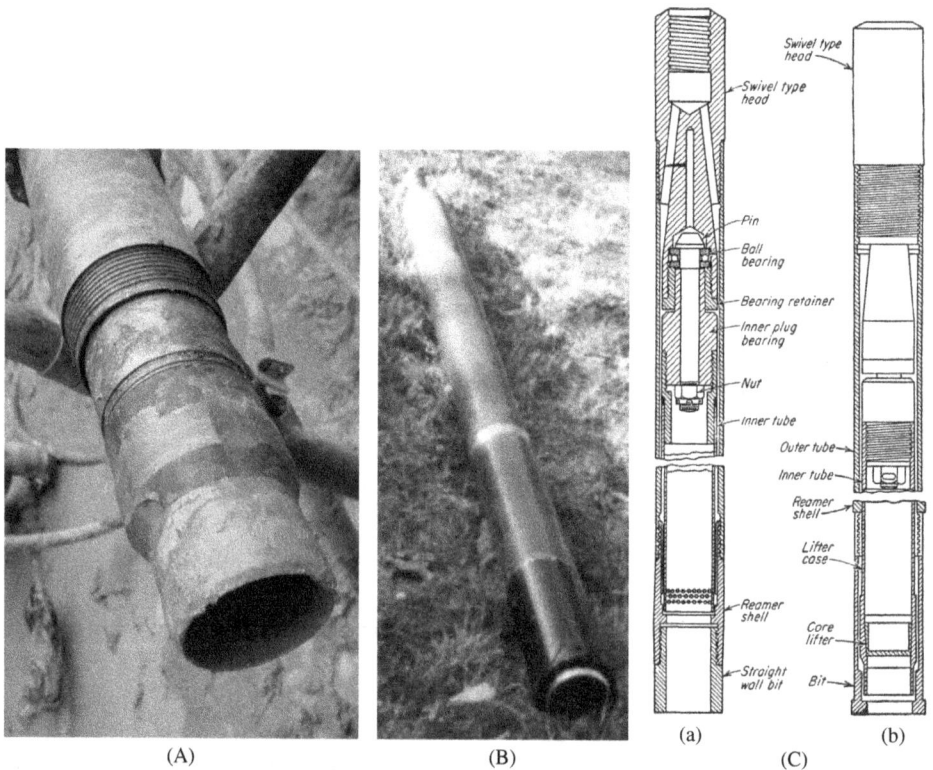

Figure 7.2   Photograph of (A) Mazier core barrel; (B) Double-tube core barrel; (C) Triple-tube core barrel.

Source: Courtesy of Chu (2002) and Jay Mfg. Co.

Table 7.4   Sizes of rotary core barrels.

Ref. on Borehole Record	Core Barrel Design	Nominal Diameter of Core (mm)	Nominal Diameter of Hole (mm)
Sizes given in BS 4019-1:1974			
B	BWF, BWG or BWM	42.0	60.0
N	NFW, NWG, NWM	54.5	76.0
	HWF or HWG (HWAF)	76.0 (70.5)	99.0
P	PWF	92.0	121.0
S	SWF	112.5	146.0
U	UWF	140.0	175.0
Z	ZWF	165.0	200.0
Miscellaneous sizes			
TBX	TBX (thin wall)	45.0	60.0
TNX	TNX (thin wall)	61.0	76.0
NQ3	NQ3 (triple tube wireline)	45.0	76.0
NQ	NQ (wireline)	47.5	76.0
HQ	HQ (wireline)	61.0	99.0
PQ	PQ (wireline	83.0	121.0
NMLC	NMLC (triple tube)	52.0	76.0
HMLC	HMLC (triple tube)	63.5	99.0
Mazier		75.0	101.0
		108.0	146.0
Metric Sizes			
T2 56	T2 56	42.0	56.0
TT 56	TT 56	45.5	56.0
T6 66	T6 66	47.0	66.0
T2 66	T2 66	52.0	66.0
T6 76	T6 76	57.0	76.0
T2 76	T2 76	672.0	76.0
T6 86	T6 86	67.0	86.0
T2 86	T2 86	72.0	86.0
T6 101	T6 101	79.0	101.0
T2 101	T2 101	84.0	101.0
T6 116	T6 116	93.0	116.0
T6 131	T6 131	106.0	131.0
Geobor S	SK6L	102.0	146.0
T6 146	T6 146	123.0	146.0

KEY

TT is extra thinwall water barrel

T2 is water barrel

T6 is mud water

SK is waterline barrel

sub-surface structures such as foundations and buried structures. Corrosion ability to steel and concrete is an important parameter that is required to be checked.

Groundwater samples are collected from either open holes or case holes after drilling or from the installed water standpipes or monitoring wells. Groundwater samples are collected with a suitable jar and the temperature is maintained by storing the jar in a cool box and sending it to a groundwater chemistry laboratory within 24 hours of sampling. Change of custody is also required in the process of delivering the sample. Figure 7.3 shows groundwater sampling in progress using a jar and container. Filtration, preservation, documentation, and quality assurance of groundwater sampling can be found in ASTM D 6564, D 6517, D 6089, and D 7069, respectively.

7.3 Sampling Interval

Many clients and geotechnical engineers prefer continuous sampling throughout the drilling. Continuous sampling is carried out with an interval of one meter and additional samples are usually collected whenever formation changes are encountered. In some cases, sampling intervals are based on either drill rod length or sampler length. In Ontario, Canada, it is common to collect samples as close as every 0.75 m until stress influence depth of the proposed foundation, and every 1.5 m until a depth of 20 m and every 3 m beyond the depth of 20 m.

7.4 Soil and Rock Samplers

There are various types of samplers which can collect disturbed and undisturbed samples during drilling investigation. As explained earlier, disturbed samples can

Figure 7.3   Groundwater sampling in progress using a jar and container.

be collected from excavated soils from the test pit and trenches as well as from auger returns and suspended soils retained from drilling fluid returns. In addition, disturbed samples can be collected from the recovered soils during SPT testing. Disturbed soils sample can also be collected using window sampling.

Undisturbed samples can be collected from test pits and trenches and during drilling investigation, using various types of undisturbed soil samplers. These will be discussed in detail in the following sections.

7.4.1 *Split spoon sampler*

A split spoon sampler is used to obtain the Standard Penetration blow counts in the granular and firm to stiff cohesive soils. The recovery of samples is reasonably good in stiff clay and clayey silt or clayey sand due to the existence of cohesion. However, the recovery of samples is not very good in clean and loose sand as well as soft to very soft clay or loose silt. Samples are normally disturbed due to the forces applied during driving of the sampler, and generation of excess pore pressure occurred in the soil masses.

The sampler is 600 mm in length, with an outside diameter of 51 mm and inside diameter of 35 mm. It has a thickness of 16 mm which also causes significant disturbances to the samples collected. After the sample is retrieved from the borehole, samples can be obtained by splitting the two halves of the sampler. During the process of SPT testing, the sampler is driven into the bottom of borehole with the SPT hammer with a 63.5 kg weight dropped from a 750 mm height. Total driving length is 450 mm including the initial drive of 150 mm. Details of the SPT test are described in Section 4.6.1 and can also be found in ASTM D 1586-18 and BS 5930 (1999) and BS 1377 — Part 9 — 1999. Figure 7.4 shows samples retrieved from the split spoon sampler.

Figure 7.4   Retrieved sample from a split spoon sampler.

7.4.2 *Window sampler*

This sampler is suited for dry cohesive soils and possibly to take samples up to an 8 meter depth. The sampler itself is 1000 mm which can be screwed into the driving or pushing rod. A cutting shoe can also be screwed onto the sampler where hard driving is required. Various sizes of diameter such as 36, 50, 60, and 80 mm are available in the market. The sampler is driven or pushed into the ground by a rig-operated hammer and extracted applying hydraulic forces. The sampler has a side window slot to receive the soils which enter into the sampler. With this approach of driving continuously, continuous disturbed soil samples can be obtained. Collecting soft and loose soils can be achieved with static pushing, whereas collecting firm to stiff and dense soils requires applying driving forces. Figure 7.5 shows sampling in progress with a window sampler.

7.4.3 *U100 sampler*

The U100 sampler, which is similar to the 100-mm Shelby tube in North America, is a thin-wall sampler which has a small area ratio of approximately 27%. An inside clearance of 1.4% is provided due to tapering of the 30 degree cutting edge. The thin wall with a small area ratio and sharp cutting edge promotes minimum disturbance, and therefore nearly undisturbed samples can be

Figure 7.5   Sampling process using a window sampler.
Source: ADP Group Ltd.

obtained using the U100 sampler in the cohesive soils. The sample diameter is normally 100 to 106 mm with a length of 457 mm. Collected samples are sealed with wax at the two ends of the tube to maintain the *in-situ* moisture content. For long-term storage, wax is usually mixed with petrolatum (i.e., Vaseline) to prevent drying and cracking of the seal. Figure 7.6 shows a typical thin-walled sampler. The sampler is attached to the drill rod and pushed into the formation applying static forces.

7.4.4 *Piston sampler*

The piston sampler uses the same thin-wall tube as the U100 and is available in 75 mm diameter and 100 mm diameter. Figure 7.7 shows two types of piston samplers. The piston sampler head is equipped with a piston within a tube which is initially positioned at the bottom of the sample tube before pushing the sample tube into the soil (Figure 7.8(a)). This prevents soils from entering into the sampler before the sampling process is commenced. When the sample tube is pushed into the soil, the piston is progressively retrieved to the top of the tube when soil mass enters into the tube (Figure 7.8(b)). This process creates a vacuum within the

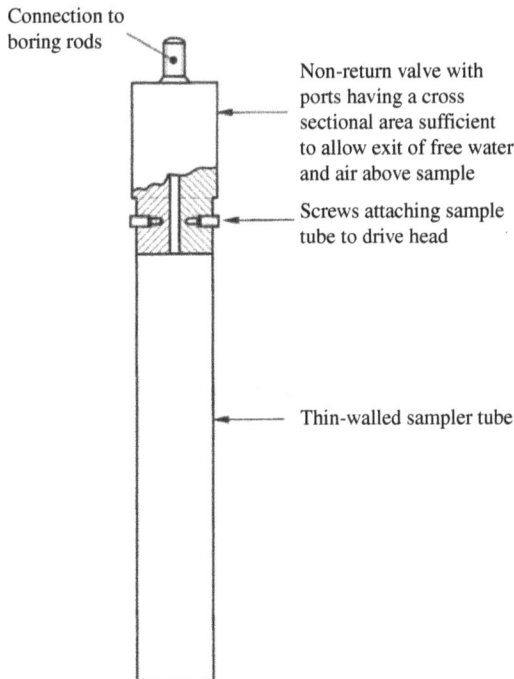

Connection to
boring rods

Non-return valve with
ports having a cross
sectional area sufficient
to allow exit of free water
and air above sample

Screws attaching sample
tube to drive head

Thin-walled sampler tube

Figure 7.6   A typical thin-walled sampler.
Source: BS5930 1999.

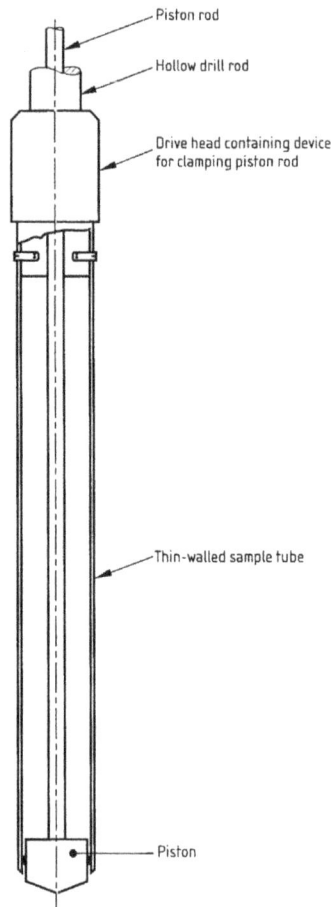

Figure 7.7   Piston samplers.
Source: BS 5930 1999.

tube, and therefore this suction pressure helps hold the soil with minimum disturbance. Therefore, a piston sampler provides a better-quality sample compared to
any other sampler.

7.4.5 *Large-diameter samplers*

It has been known that larger-dimension samplers provide better-quality samples
for laboratory testing as they have better ability to maintain the *in-situ* condition.
Many people have tried collecting larger-diameter samples from the boreholes in
many big projects which dealt with soft clay. The Changi Land Reclamation project in Singapore had collected certain numbers of 180-mm large-diameter samples from both upper and lower marine clay which is soft to firm in consistency.

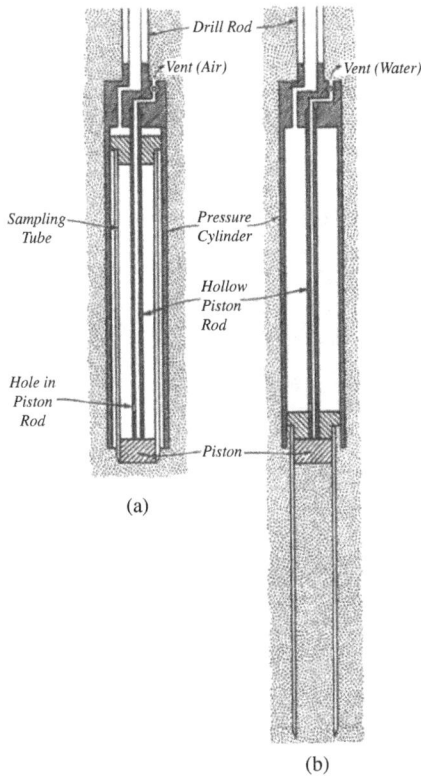

Figure 7.8 Piston sampler of hydraulically operated types: (a) lowered to bottom of drill hole, drill rod clamped in fixed position at ground surface. (b) sampling tubes after being forced into soil by water supplied through drill rod.

Figure 7.9 shows extruding of a 180-mm sample in the on-site laboratory at the Changi Land Reclamation project. This sample tube has an area ratio of 30. In Quebec, Canada, two larger-diameter samplers were developed by two universities in Quebec to collect the sensitive Leda clay which is soft. These samplers are commercially available nowadays.

7.4.5.1 *Laval sampler*

A new Laval sampler has been developed by a design group from Laval University, Canada. The internal diameter of the sample tube is 208.3 mm and the external diameter is 216.3 mm. It has a thickness of 4 mm. It provides a sample tube with an area ratio of 52. The length of the tube is 660 mm. The tube is fairly rigid. The sampler head has four screws, and the general shape of the sampler head is shown in Figure 7.10. It is made with steel and has a 30-mm hole in the center connected to the lateral opening that allows the mud fluid to flow out of

Figure 7.9   Sample extrusion process of a large-diameter sample.

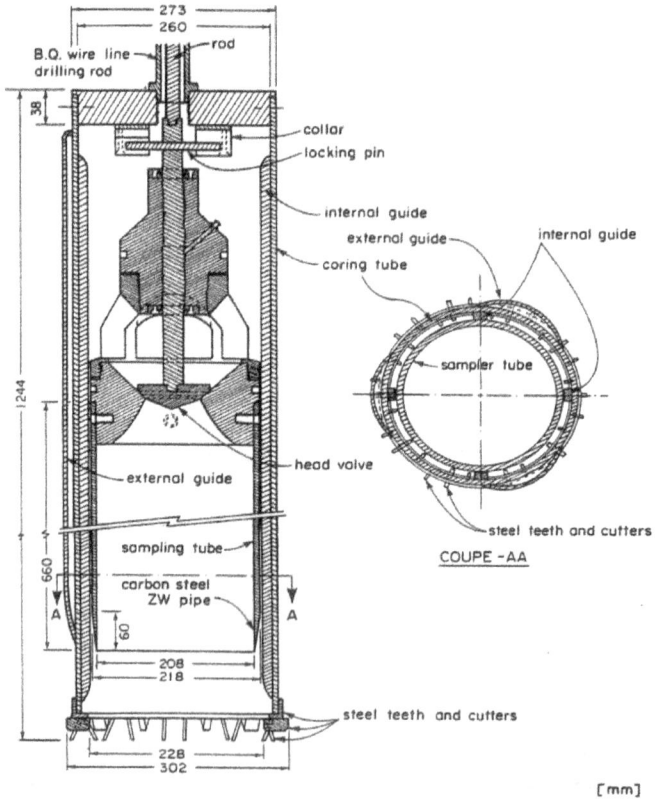

Figure 7.10   Technical drawing of a Laval Sampler (200 mm diameter).
Source: La Rochele *et al.* (1981).

the sample tube and into the coring tube when it is pushed down into the soil. The hole can be plugged by a 35-mm steel ball, which prevents mud fluid from flowing back into the sampler head and pushing the sample out of the tube when it is lifted to the surface (Oka *et al.*, 1996). The preparation of the borehole for a Laval sampler is either done by the previous sampling or by means of a solid auger with a diameter of 400 mm. The borehole can be supported by mud or be cased down to the sampling level. The sampler assembly is lowered into the borehole with the sampler hooked up inside the coring tube and with the head valve open; the mud can then flow freely through the sampler. When the lower edge of the coring tube reaches the bottom of the borehole, the coring tube is held fixed from the surface and the sampler is unhooked by pulling up and turning the central rod slightly. As the tube sampler is pushed down into the soil by a continuous thrust, the mud flows out of the tube through the head valve of the sampler. To make sure that no pressure is applied on the soil sample, the movement of the sampler is stopped when the head of the sampler has reached an elevation of approximately 50 mm above the top of the sample. The head valve is then closed, and the coring operation is carried out by rotating the coring tube, while at the same time injecting the bentonite mud under pressure. This flows through the drill rod down between the sampling and coring tubes, around the lower remoulding ring, and up outside the coring tube into the borehole. The injection of mud aims to wash the remoulded clay out of the teeth and cutters of the remoulding ring. When the coring ring has reached a depth of approximately 20 mm below the edge of the sampler, the coring is stopped and the sampler rotated through 90, pulled up gently, and hooked back on the collar of the coring tube ready to be retrieved from the borehole. Figures 7.10 and 7.11 show a Laval sampler and Figure 7.12 shows the process of sample collection. Figure 7.13 shows the view of the sample tube after collection.

The soil samples are extruded immediately after sampling in the field. They are cut out with a wire in slices 130 mm or 200 mm high, depending on the type of tests to be carried out. The slices are put on a waxed plywood board, wrapped in special paper, sandwiched between layers of a paraffin wax and Vaseline mixture, and are then ready to be transported and stored. Figures 7.14 and 7.15 show the preparation of a Laval sampler before sampling and the sample extruding process from the Laval sampler, respectively.

7.4.5.2 *Sherbrooke sampler*

The Sherbrooke sampler was designed by Levebvre and Poulin (1979). The sampler is essentially a down-borehole block sampler. The feature of the sampler is shown in Figure 7.16. The sampler needs a borehole of about 400 mm diameter, which is best cleaned using a flat-bottomed auger, in order to reduce disturbance

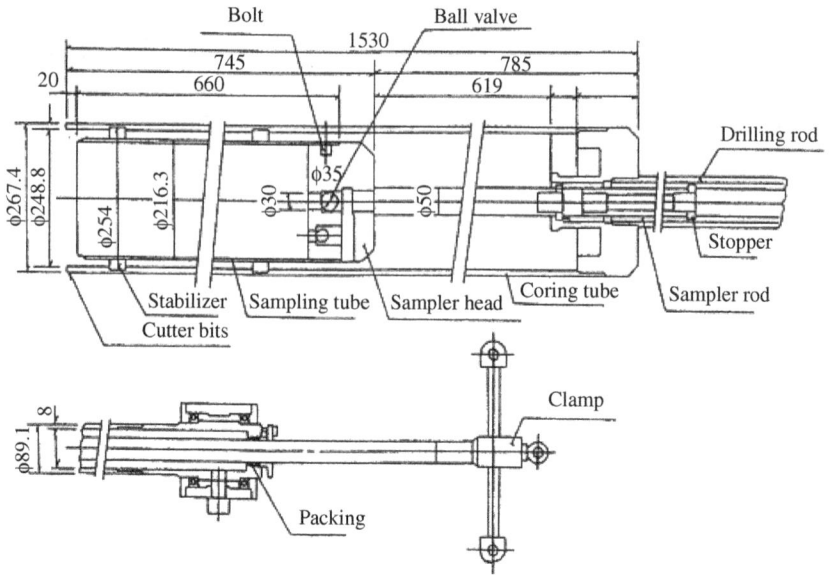

Figure 7.11   Technical drawing of a Laval Sampler.

Source: Oka *et al.* (1996).

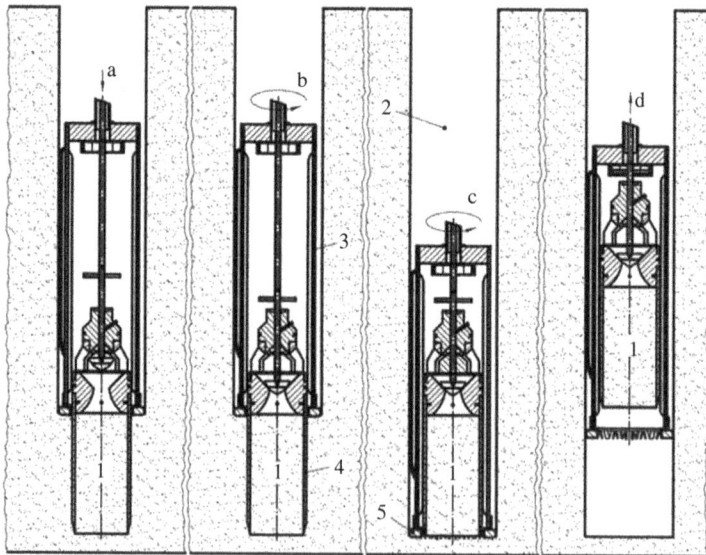

Figure 7.12   Process of sample collection using a Laval Sampler.

Notes:

1. Sample; 2. borehole 300 mm in diameter; 3. coring tube; 4. sampling tube; 5. cutting teeth; a. The tube sampler is pushed down; b. The head valve is closed by screwing the inner string of rods; c. The coring operation is carried out by rotating the coring tube; d. The sampling tube is hooked back on the collar of the coring tube and the sampler is retrieved from the borehole.

Source: https://Geotechnicaldesign.info.

and minimize the amount of disturbed material left in the base of the hole before sampling. The borehole can be supported by mud or be cased down to the sampling level. The hole is usually kept full of drilling fluid. The sampler is connected to an ordinary drill rod system and lowered to the base of the hole. The drill rod provides rotation of the sampler at about 5 r/min during the carving. The sampler is rotated, either by hand or using a small electric motor, at about 5 r/m. A cylinder of soil, about 250 mm in diameter, is carved out by three circumferential blades,

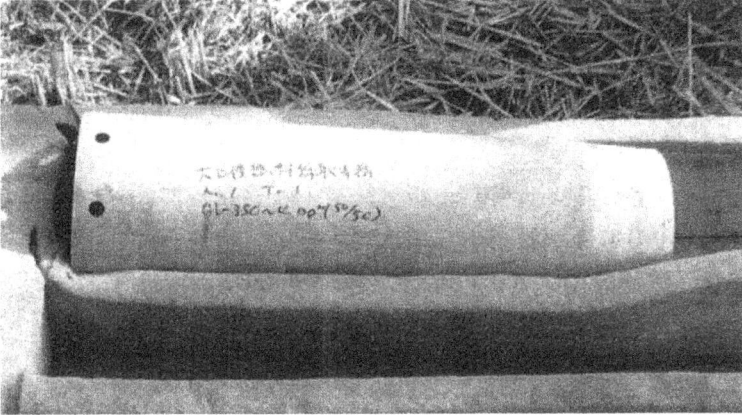

Figure 7.13   Collected sample using a Laval Sampler.
Source: Oka *et al.* (1996).

Figure 7.14   Laval sampler preparation prior to sampling.
Source: Chu 2002.

Figure 7.15   Sample extruding process of a Laval Sampler.
Source: Chu 2002.

spaced at 120 degrees. They make a slot about 50 mm wide and are fed with bentonite or water to help clear the cuttings. The time taken to obtain a sample obviously depends upon ground conditions, but may be about 30–40min. After carving out a cylinder about 350 mm high, the operator pulls a pin, and the blades (which are spring-mounted) gradually rotate under the base of the sample, as rotation is continued. Closure of the blades separates the sample from the underlying soil, and the sample is then lifted to the surface with a block and tackle. Lifting takes place very slowly for the first 0.5 m, in order to avoid suction at the base of the sample. Figure 7.17 shows the process of sampling using the Sherbrooke sampler. Figure 7.18 shows the sampling in progress and Figure 7.19 shows sample extrusion in progress. The sample is coated with layers of paraffin wax and may be placed in a container packed with damp sawdust or other suitable material. The complete process takes about 3 h, including preparation for shipment. Tests by Lefebvre and Poulin have shown that this sampler is capable of obtaining soil of comparable quality to that produced by block sampling in the sensitive clays of eastern Canada. Where the highest-quality samples are required for testing of soft or sensitive clays, this apparatus provides the best method of obtaining undisturbed samples from depth.

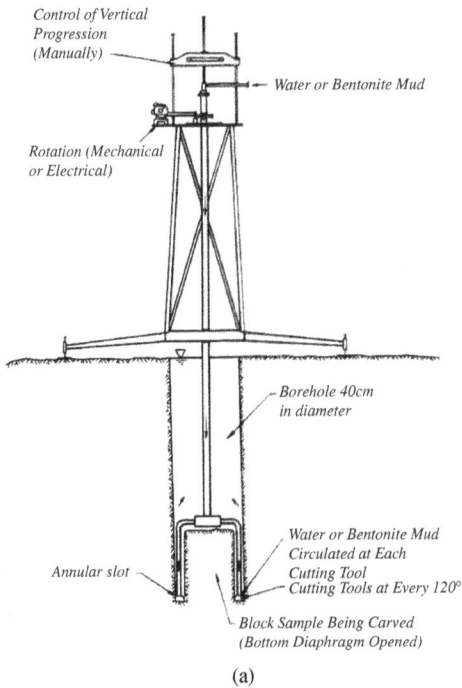

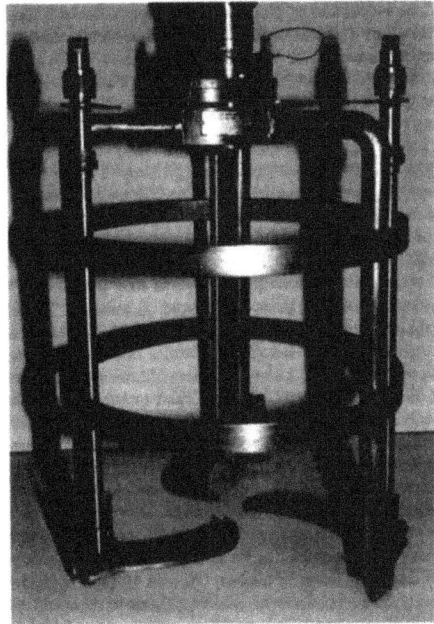

Figure 7.16 (a) Diagrammatic sketch of the Sherbrooke apparatus for carving block sample beneath drill hole. Reproduced with permission from Lefebvre and Poulin (1979). (b) Photograph of a Sherbrooke sampler (after Terzaghi *et al.* 1996).

7.5 Packaging and Sample Identifications

7.5.1 *Sample preservation and packaging*

During investigation borehole drilling, the representative soil samples are collected for visual classification purposes and further geotechnical laboratory testing purposes. Depending upon the type of soil encountered during the drilling of the borehole, suitable types of samples are collected. Samples are required to be packed, preserved, transported, stored, and extruded using a suitable method which could minimize the post-sampling disturbances. Undisturbed samples should also be preserved to maintain the *in-situ* state as much as possible.

Suitable types of sample containers are used depending on the class of the sample, local climate conditions, transporting mode, and distance.

7.5.2 *Disturbed sample preservation and packaging*

Disturbed samples should be packed into an airtight plastic bag preferably in two layers to prevent moisture loss. SPT samples should be placed into a plastic

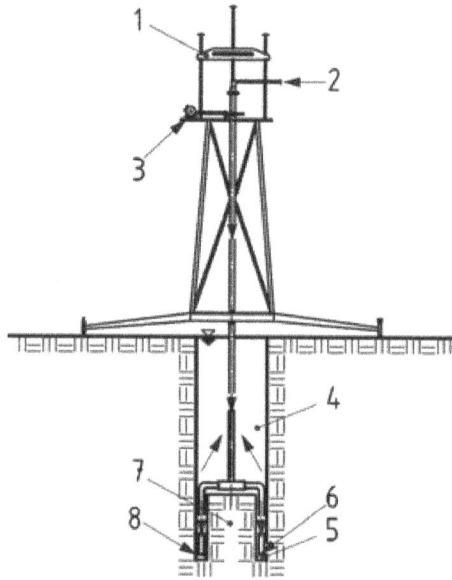

Figure 7.17   Example of sampling from borehole bottom using a large sampler (Sherbrooke block sampler).

Notes:
1. control of vertical progression (manually); 2. annulus slot; 3. rotation (mechanic or electrical); 4. water or bentonite mud; 5. borehole 400 mm in diameter; 6. water circulating at each leg; 7. sample being carved (bottom diaphragm opened); 8. cutting tools at every 120.
Source: https://geotechnicaldesign.info.

Figure 7.18   Sampling in process using a Sherbrooke sampler.
Source: Chu 2002.

container with a diameter slightly larger than the inner diameter of the SPT sampler. All plastic containers storing the samples retrieved from the single SPT split spoon should be packed into one big plastic bag.

7.5.3 *Undisturbed sample preservation and packaging*

Undisturbed samples collected using the U100 or piston samples are sealed at the top and bottom of the sample tube with microcrystalline wax. Where longer storage is required, Vaseline is mixed with wax to prevent cracking. In many cases, a thin metal disc of a diameter slightly less than the sample tube diameter is used together with a cap. A cap made of plastic, rubber, or metal is used to cap the opening ends of the sample tube after sealing with the wax. The ends of the tube are also tapped with waterproof duct tape. Figure 7.20 shows the application of wax in progress.

Undisturbed samples collected as block samples from the test pit and trenches should be wrapped in suitable plastic film or/and aluminum foil and coated with several layers of wax.

The waxed block samples should be placed into a wooden box or any other solid box such as plastic or metal which fits the size of the samples.

Details on preservation of soil samples can be found in ASTM D 4220-14.

Figure 7.19   Extruding process of sample from a Sherbrooke Sampler.
Source: Chu 2002.

Figure 7.20   Process of preserving samples using wax.
Source: Courtesy of Jian Chu.

7.5.4 *Preservation of water samples*

The water sample containers are generally kept in a dark, filled, and thermally insulated or refrigerated room, without any contact with materials that could affect the water quality. Field preservation of groundwater samples should comply with ASTM D 6517 and D 7069. The samples should be transported to the laboratory daily.

7.6 Labeling of Samples

All samples should be immediately numbered, documented, and labeled after sampling and sealing.

The label should show the following information:

- project name and number;
- type of investigation such as test pit, trench, and borehole;
- date of sampling;
- sample identification number and investigation borehole number;
- type of sample;
- total recover length;
- depth of the sample from reference level.

The top and bottom of the samples should be marked to identify the upper and lower end of the sample. The label should also indicate the soil and rock type.

7.7 Transportation of Samples

When samples are transported, national laws or safety regulations should be complied with if samples are known or suspected to contain hazardous material. A traceability record should accompany the shipment so that the possession of the sample is traceable from collection to shipment to receiving at the laboratory. In each stage of transferring the samples, the person(s) receiving the samples should sign and record the date and time.

Samples should be handled in the same orientation in which they were sampled, including during transportation or shipping (Figure 7.21). Sample tubes and containers are placed on or wrapped with rubber foam to protect against vibration and shock. Where required, sample containers and tubes are insulated with either granules or foam to be able to resist the temperature change of the soil or to prevent the samples from freezing. Samples should only be placed in solid boxes into which they fit snugly, thereby preventing bumping, rolling, dropping, etc. The cushioning material (sawdust, rubber, polystyrene, urethane foam, or material with similar resiliency) should completely encase the samples in such a way that they are not disturbed during transport.

Details of transportation can be found in ASTM D 3213-13 and D 4220-14.

Figure 7.21   Samples prepared to be transported to testing facilities.
Source: Courtesy of Jian Chu.

7.8 Sample Storage

Every soil and rock sample should be protected at all times from direct sunlight, heat, frost, and rain. Samples should be stored in a moisture control storage room preferably in the refrigeration unit, capable of maintaining *in-situ* temperature.

Storage time should be minimal as some properties may change for some types of soil within hours or days of sampling. Storage should not be allowed to adversely affect the soil properties to be measured.

7.9 Sample Extrusion

To prepare for the testing of soil samples and/or storage, samples are required to be extruded from the sample tubes. An extruder normally pushes out the sample gently with the use of static force. Most extruders use hydraulic pressure, whereas some use mechanical screwing. There are two types of extruders such as horizontal and vertical extruders. As ASTM D 4220 recommended that the samples be handled the way they were sampled, it is recommended that tube samples should be extruded with a vertical extruder. Figure 7.22 show vertical and horizontal extruders.

7.10 Sample Disturbances

Despite the best available methods being utilized to collect undisturbed samples, the collected samples have a certain degree of disturbance due to many reasons.

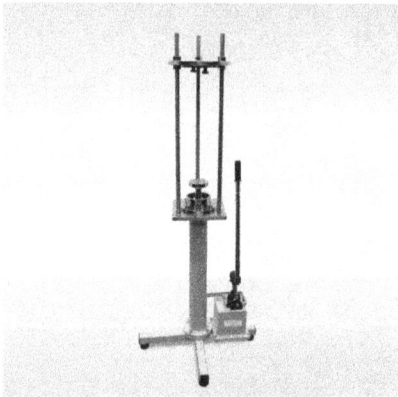
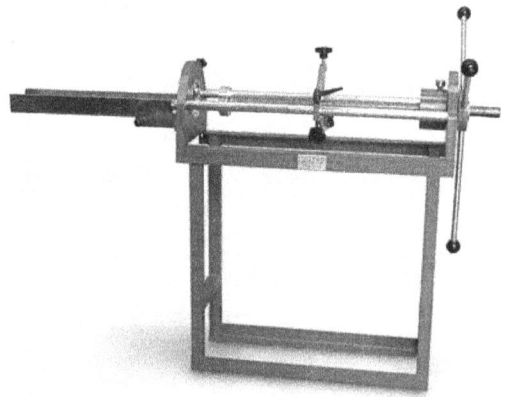

(a)   (b)

Figure 7.22   Various types of sample extrusion equipment: (a) Vertical extruder using hydraulic jack; (b) Horizontal extruder using screw.

Source: (a) UTEST (b) MATEST.

Samples suffer from disturbances due to poor drilling operations, forces being applied during direct pushing for sampling, mechanical distortion of soil during penetration of the sample tube, some displacements occurring during penetration, suction generated during retrieval of the sample tube, water content redistribution in the tube, disturbances due to rough handling and transportation, drying and wetting, and release of total and effective stress during and after sampling. The last one is unavoidable. Figure 7.23 shows various types of disturbances during various events and their stress passes. Hight (1986) listed the change in effective stresses during the sampling process as shown in Figure 7.24. Baligh (1985) and Baligh *et al.* (1987) demonstrated that strain occurs during sampling as shown in Figure 7.25 in which greater strain occurs in the sample tube with a smaller ratio of the tube diameter to the wall thickness. Therefore, larger-diameter samplers with a thinner wall can obtain samples with less disturbance. Therefore, large-diameter samplers such as the Laval and Sherbrooke samplers are frequently recommended to obtain good-quality soft soil samples. Figure 7.26 shows a change in pore water pressure in normally consolidated soils and overconsolidated soils after sample collection. A sampler with sharp tip can also provide better-quality samples (Figure 7.27). Figure 7.28 shows a comparison of results obtained from Laval and piston samples for unconsolidated, undrained triaxial compression tests. It can be clearly seen that a larger-diameter sample provides higher strength due to less disturbances.

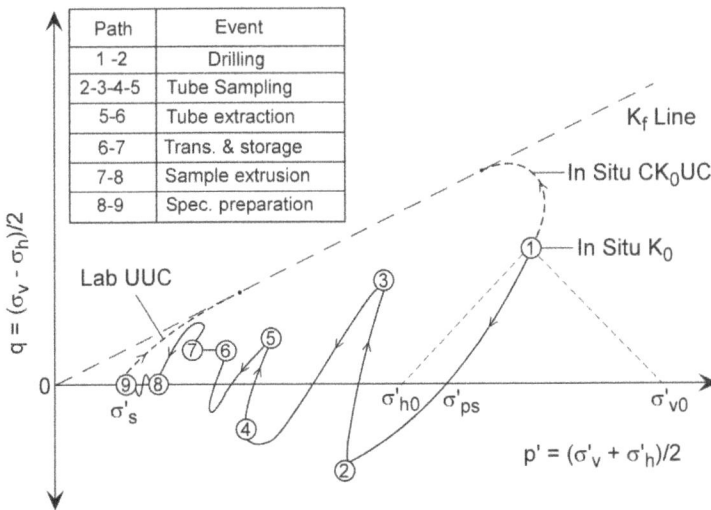

Figure 7.23   Various types of disturbances during various events and their stress passes.
Source: After Ladd (2003).

Source of disturbance	Potential change in p'	
	Decrease in p'	Increase in p'

Borehole instability
Tube sampling
   Centerline strains
   Peripheral strains
Sample transport
Sample storage
   Drying
   Gas diffusion
Sample extrusion
Specimen preparation
   Trimming disturbance
   Drying
Specimen installation
   Membrane
     Pre-soaked
     Unsoaked
   Filter-paper drains
   Specimen seating
     Coarse porous stones
     Fine porous stones
Cavitation in sand/silt laminae
   Entry of borehole fluid
   Unsaturated laminae

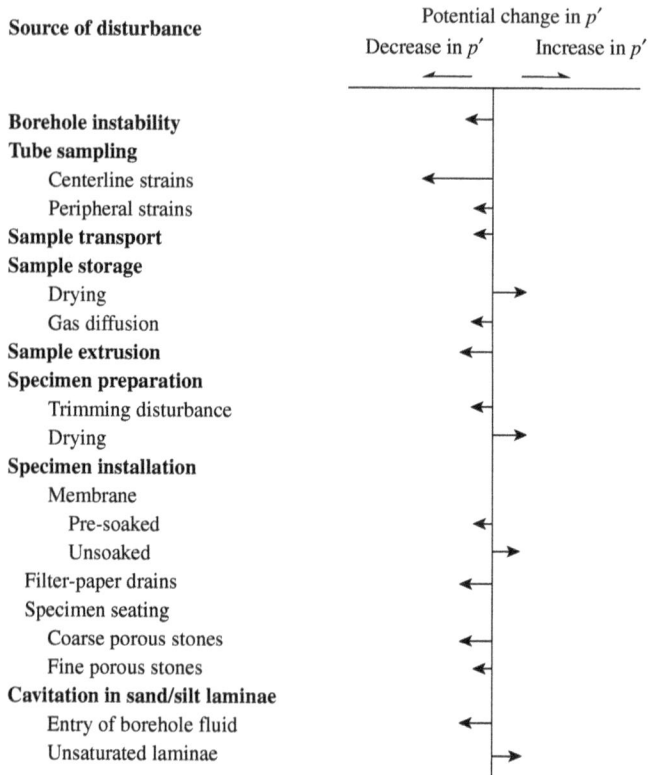

Figure 7.24   Factors influencing the mean effective stress in specimens of soft clay.
Source: Hight (1986).

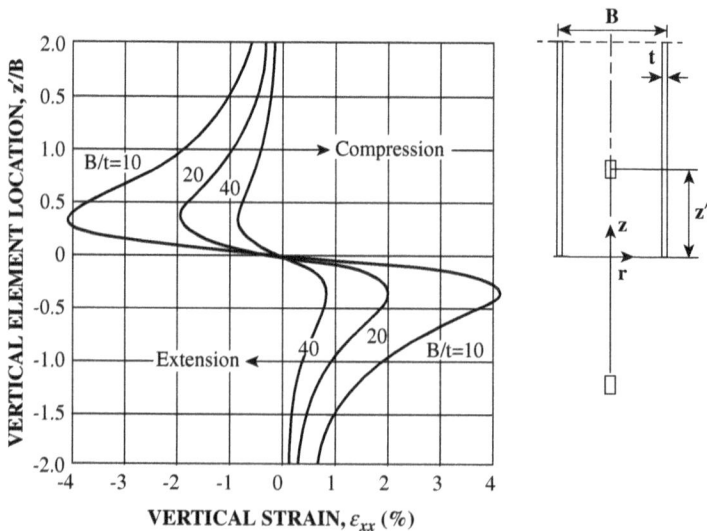

Figure 7.25   Strain-induced sampling.
Source: Baligh (1985).

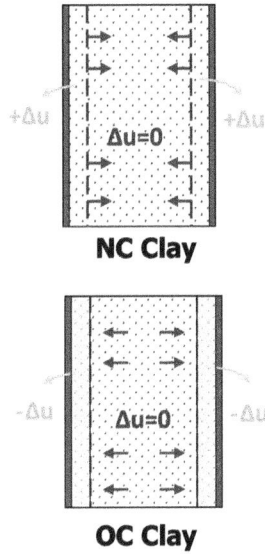

Figure 7.26 Pore water pressure changes in normally consolidated soil and overconsolidated soil after sample collection.

Source: Courtesy of Jian Chu.

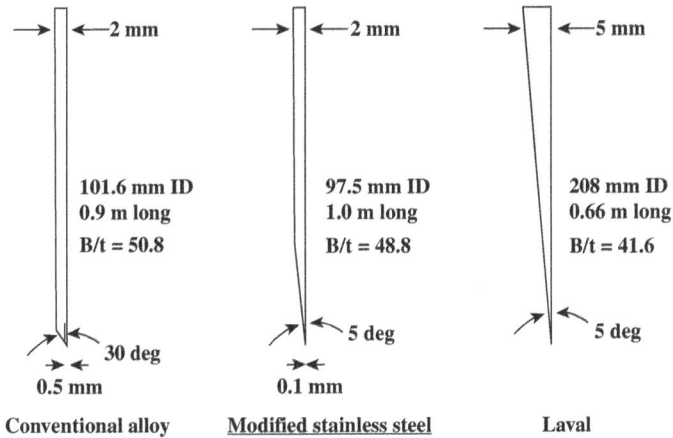

Figure 7.27 Comparison of Sampler tips for large-diameter samplers.

Source: D. Hight.

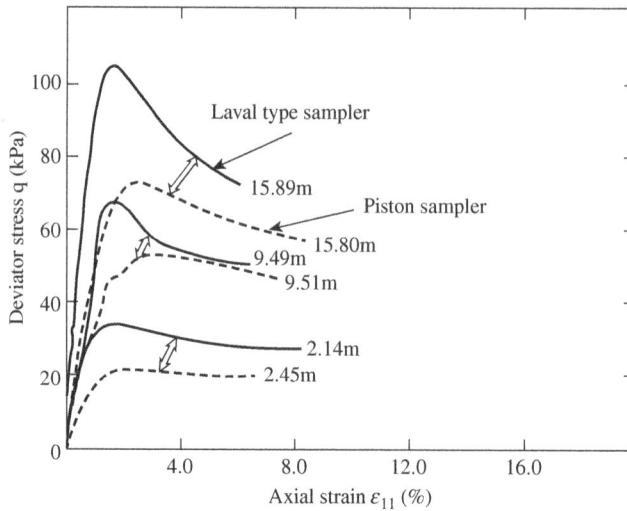

Figure 7.28   Comparison of results obtained from Laval and piston samples for unconsolidated, undrained triaxial compression tests.

Source: Hight *et al.* (1992).

Table 7.5   Size of particle and mass required for assessment sieving.

Largest Size of Particle Present in Substantial Proportion, i.e., More than 10%	Minimum Mass of Sample to be Taken for Assessment Sieving
(mm)	(kg)
< 75 but > 20	15
< 20 but > 2	2
< 2	0.1

Source: BS 1377-1:1990.

7.11 Sample Preparation

Before preparation of samples for laboratory testing, an assessment has to be made of whether it is cohesive or granular soil and their respective grain sizes. Depending upon the grain size, a suitable mass of soil should be prepared for testing. The following table suggests a suitable mass of soil based on grain size. British Standard BS 1377 (1990) recommended the required masses of the soils for assessment sieving based on their grain sizes as shown in Table 7.5. Table 7.2 also shows soil masses required for various types of classification and compaction tests. Where required, the

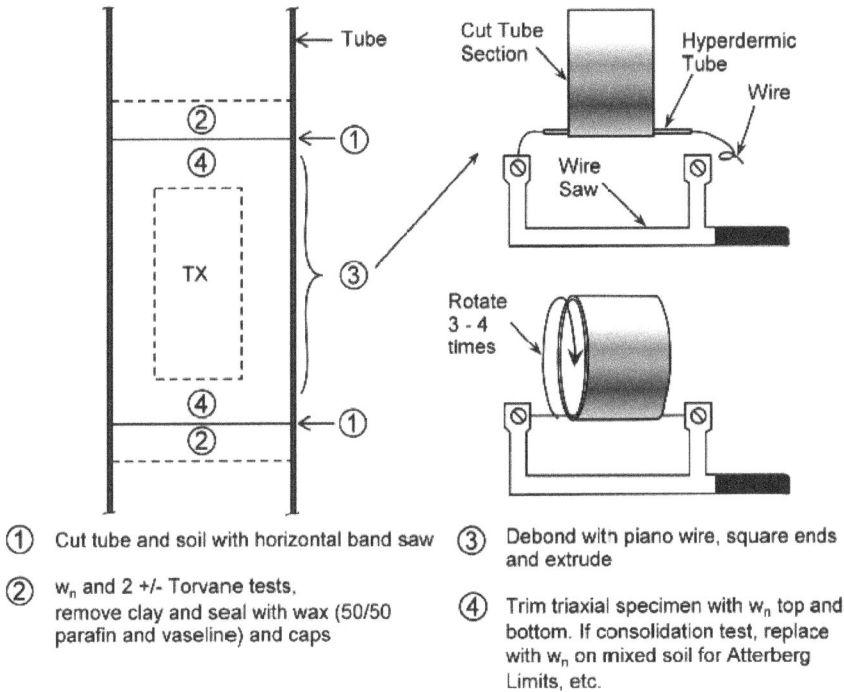

Figure 7.29   Sample preparation using a soil lathe machine.
Source: After Ladd (2003).

sample for classification tests should be dried in the laboratory to remove all moisture or dried to the level of moisture required. Oven drying is normally carried out.
Any aggregations occurring during drying should be broken down without crushing
the grains. Undisturbed samples extruded from the sample tube should be cut to the
required thicknesses using the wire saw on the soil lathe machine (Figure 7.29).
Sample preparation procedures recommended in ASTM for each type of test should
be complied with to minimize sample disturbances during preparation.

Chapter 8

Geotechnical Instrumentation

During the ground investigation for geotechnical purposes, some types of geotechnical instruments are planned and installed in order to understand the conditions of groundwater and its fluctuations, as well as to obtain the baseline conditions of the ground, the existing infrastructure locations, and elevations, which are required to be protected during the development of infrastructures. These geotechnical instruments are as follows:

- Settlement plate
- Settlement marker/point
- Deep Reference Point
- Piezometers
- Water Standpipe
- Inclinometer

8.1 Settlement Plate

A settlement plate could be installed on the ground with a certain amount of protection against damage during construction. The riser rod should be protected with a friction reducer sleeve pipe. In many cases, the settlement riser rod and protective casing could be flush finished with the existing ground and covered with a protective cap. A typical design of a settlement plate is shown in Figure 8.1. These surface settlement plates can measure the total settlement of the ground. Figure 8.2 shows the photographic features of a settlement plate and Figure 8.3 shows a settlement plate installation in progress. Generally, the level of the top of the settlement rod is measured by survey methods and the settlement of the ground is monitored. Table 8.1 shows the settlement monitoring data and Figure 8.4 shows a typical monitoring record in graphical presentation. Generally, settlement records are shown together with a soil profile and construction activities as shown in Figure 8.4.

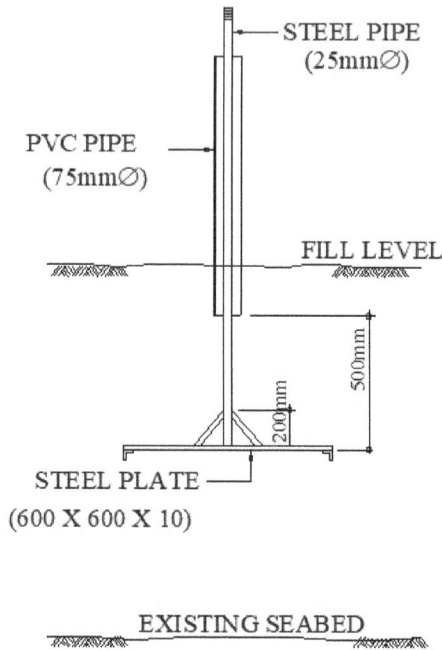

Note: Measurement shown are in mm

Figure 8.1   Typical design of a surface settlement plate.
Source: Bo and Choa (2004).

8.2 Settlement Marker/Point

Settlement markers are usually marked on the existing infrastructures of interest, which are required to maintain their stability and positions. In most cases, reflective markers are used to mark the position. In some cases, a steel rod is embedded into the concrete structure to mark the locations. Both the coordinates and elevations are obtained as baseline positions and elevations, and then the movement of the structure over time is measured and monitored. Reporting of the position and the elevations is done in the same manner as the reporting for settlement monitoring data.

8.3 Deep Reference Point

Deep reference points are installed to be used as a benchmark for survey purposes. A normal benchmark provided by the surveyor may not provide the required accuracy if the benchmark is far from the site, especially when it is installed on a pile which is driven through unstable settling ground or when it sits on the formation above a groundwater aquifer which is being exploited. In the former case, a benchmark could settle due to the down drag on the pile and in the latter case, it could

Figure 8.2   Photographic feature of a settlement plate.
Source: Bo and Choa (2004).

move down due to land subsidence caused by groundwater extraction. Unstable conditions could occur due to lowering of groundwater and seasonal thermal effects.

Therefore, a few deep reference points should be installed at site that should be anchored on the bedrock or competent formation. It should be installed with a necessary negative friction reducer. A typical design of a deep reference point is shown in Figure 8.5. In some projects, there were some records of data showing heaving of a deep reference point relative to a benchmark when checked some years after installation. Actually, it is not heaving up of a deep reference point but it is showing instead the settlement of the benchmark compared to the deep reference point a few years after installation.

8.4 Piezometers

Three types of piezometers are usually used in construction projects:

- Pneumatic Piezometer
- Vibrating Wire Piezometer
- Casagrande Open Type Piezometer

Figure 8.3   Installation of a settlement plate in progress.
Source: Bo and Choa (2004).

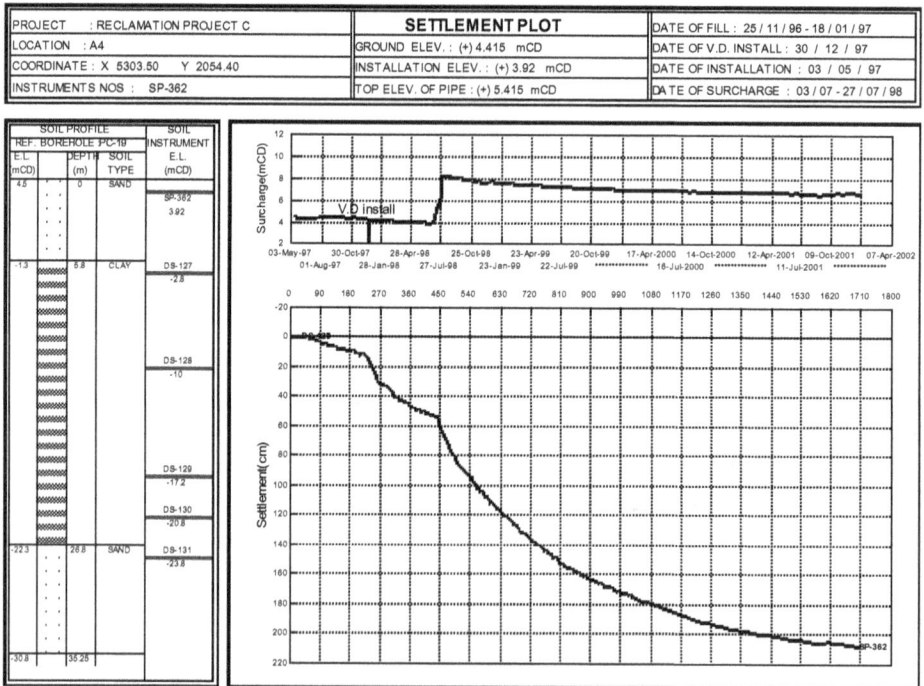

Figure 8.4   Typical settlement monitoring data in graphical presentation.

Table 8.1   Surface settlement monitoring and processed data.

PROJECT : RECLAMATION OF PROJECT C								
LOCATION : A4				SHEET : 1				
PLATE NO. : SP-362				INSTALLATION DEPTH : 0.50 m				
COORDINATE : X 5303.50 Y 2054.40				DATE OF FILL : 25 / 11 / 96 - 18 / 01 / 97				
TOP ELEV. OF PIPE : (+) 5.415 mCD				DATE OF V.D. INSTALL : 30 / 12 / 97				
INSTALLATION ELEV. : (+) 3.92 mCD				DATE OF INSTALLATION : 03 / 05 / 97				
GROUND ELEV. : (+) 4.415 mCD				DATE OF SURCHARGE : 03 / 07 - 27 / 07 / 98				

SR. NO	DATE	READING INTERVAL (day)	TOTAL READING (day)	TOP E.L. OF PIPE (mCD)	SETTLEMENT (m)	CUMMULATIVE (m)	SURCHARGE ELEV. (mCD)	HEIGHT (m)	REMARKS
1	9-May-97	0	6	5.415	0	0	4.495	0	
2	14-May-97	5	11	5.418	-0.003	-0.003	4.489	0	
3	20-May-97	14	25	5.414	0.004	0.001	4.454	0	
4	4-Jun-97	7	32	5.415	-0.001	0	4.472	0	
5	12-Jun-97	8	40	5.415	0	0	4.464	0	
6	19-Jun-97	7	47	5.416	-0.001	-0.001	4.483	0	
7	26-Jun-97	7	54	5.411	0.005	0.004	4.476	0	
8	4-Jul-97	8	62	5.407	0.004	0.008	4.48	0	
9	18-Jul-97	14	76	5.395	0.012	0.02	4.459	0	
10	25-Jul-97	7	83	5.382	0.013	0.033	4.494	0	
11	31-Jul-97	6	89	5.378	0.004	0.037	4.493	0	
12	7-Aug-97	7	96	5.368	0.01	0.047	4.524	0	
13	14-Aug-97	7	103	5.37	-0.002	0.045	4.512	0	
14	21-Aug-97	7	110	5.367	0.003	0.048	4.52	0	
15	28-Aug-97	7	117	5.362	0.005	0.053	4.513	0	
16	4-Sep-97	7	124	5.36	0.002	0.055	4.512	0	
17	11-Sep-97	7	131	5.353	0.007	0.062	4.498	0	
18	18-Sep-97	7	138	5.343	0.01	0.072	4.498	0	
19	25-Sep-97	7	15	5.338	0.005	0.077	4.506	0	
20	25-Sep-97	0	145	5.338	0	0.077	4.506	0	
21	2-Oct-97	7	162	5.336	0.002	0.079	4.495	0	
22	9-Oct-97	7	159	5.337	-0.001	0.078	4.475	0	
23	16-Oct-97	7	166	5.328	0.009	0.087	4.481	0	
24	23-Oct-97	7	173	5.322	0.006	0.093	4.483	0	
25	6-Nov-97	14	187	5.322	0	0.093	4.493	0	
26	13-Nov-97	7	194	5.313	0.009	0.102	4.488	0	
27	20-Nov-97	7	201	5.304	0.009	0.111	4.46	0	
28	27-Nov-97	7	208	5.298	0.006	0.117	4.42	0	
29	4-Dec-97	7	215	5.299	-0.001	0.116	4.412	0	
30	11-Dec-97	7	222	5.29	0.009	0.125	4.396	0	
31	18-Dec-97	7	229	5.284	0.006	0.131	4.301	0	V.D Install
32	8-Jan-98	21	250	5.248	0.109	0.24	4.29	0	Disturbed by V.D
33	15-Jan-98	7	257	5.223	0.025	0.265	4.243	0	
34	22-Jan-98	7	264	5.2	0.023	0.288	4.284	0	
35	5-Feb-98	14	278	5.177	0.023	0.311	4.243	0	
54	29-Jun-98	3	422	4.967	0.002	0.521	3.98	0	Before ext.
55	29-Jun-98	0	422	7.967	0	0.521	3.98	0	After ext.
56	16-Jul-98	17	439	7.952	0.015	0.536	5.678	1.68	
57	17-Jul-98	1	440	7.941	0.011	0.547	5.67	1.67	Before ext.
58	17-Jul-98	0	440	9.915	0	0.547	5.67	1.67	After ext.
59	20-Jul-98	3	443	9.907	0.008	0.555	5.421	1.42	
60	23-Jul-98	3	446	9.888	0.019	0.574	7.061	3.06	
61	27-Jul-98	4	450	9.834	0.054	0.628	8.335	4.34	
149	24-Jan-00	7	996	8.742	0.004	1.72	7.114	3.11	
150	31-Jan-00	7	1003	8.74	0.002	1.722	7.094	3.09	
151	14-Feb-00	14	1017	8.728	0.012	1.734	7.096	3.1	
152	21-Feb-00	7	1024	8.72	0.008	1.742	7.095	3.1	
153	28-Feb-00	7	1031	8.71	0.01	1.752	7.063	3.06	
154	6-Mar-00	7	1038	8.705	0.005	1.757	7.052	3.05	
155	20-Mar-00	14	1052	8.688	0.017	1.774	7.069	3.07	
156	3-Apr-00	14	1066	8.682	0.006	1.78	7.064	3.06	
157	17-Apr-00	14	1080	8.671	0.011	1.791	7.05	3.05	
158	2-May-00	15	1095	8.659	0.012	1.803	7.031	3.03	
159	15-May-00	13	1108	8.655	0.004	1.807	7.026	3.03	
160	29-May-00	14	1122	8.635	0.02	1.827	7.012	3.01	
161	12-Jun-00	14	1136	8.626	0.009	1.836	7.011	3.01	

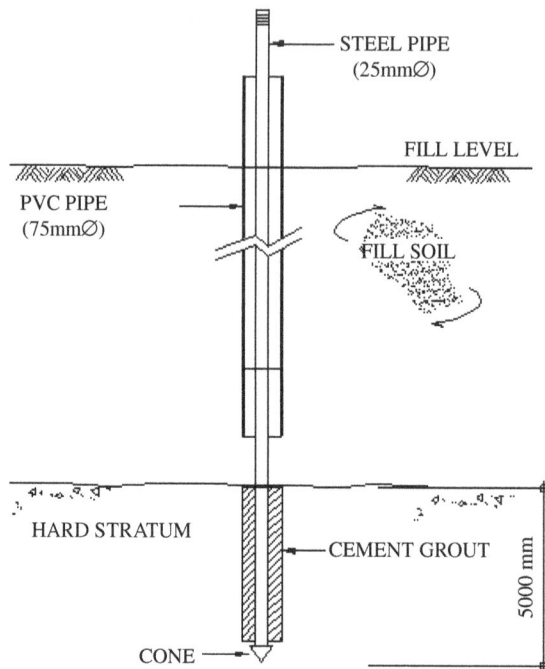

Figure 8.5   Typical design of a deep reference point.
Source: Bo and Choa (2004).

8.4.1 *Pneumatic and vibrating wire piezometers*

Piezometers are installed in a borehole. One of each piezometer should be installed in each borehole at a predetermined elevation. Pneumatic and vibrating wire piezometers should be calibrated in the local environment before installation. On-site calibration could be carried out in a large-diameter tube well and validate the pressure measured against actual water column pressure on the piezometer as a field check (Figure 8.6). In the case of a vibrating wire piezometer, the measurement is frequency which can be converted to actual water pressure using the calibration chart provided by the manufacturer. Examples of a site calibration chart and graphical presentation are shown in Table 8.2 and Figure 8.7, respectively.

Piezometers are packed in a sandbag and saturated in water at least twenty-four hours before installation. After installation in a borehole, sand should be placed again around the piezometer to the certain level above the piezometer to form an filter envelope. Then, a bentonite seal should be placed on top of the sand column. On top of the bentonite plug, the borehole should be backfilled up to the original ground level, preferably with the original soil. If not, it should be backfilled with a good mixture of bentonite cement where permeability is equivalent to or lower than the natural soil. Backfilling with sand would lead to

Figure 8.6   Site calibration in progress.
Source: Bo and Choa (2004).

SITE CALIBRATION RESULT
PNEUMATIC PIEZOMETER (PP-203)

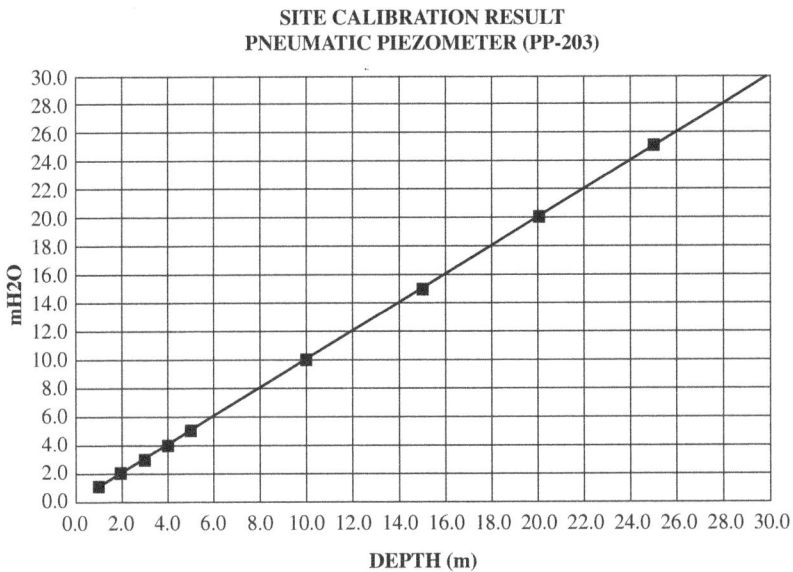

Figure 8.7   Site calibration chart for a piezometer.
Source: Bo and Choa (2004).

underestimation of the pore pressure at the measured location due to rapid dissipation of the pore pressure along the sand fill column above the piezometer. A typical installation design of a piezometer is shown in Figures 8.8 and 8.9. Figure 8.10 shows photographic features of pneumatic and vibrating wire piezometers.

Table 8.2   Typical site Calibration data.

S/N			DATE: 12-January-98		
Cluster No.: A4S-07 (A4 Area)					

1. Site Calibration Data

Depth (m)	Reading (PSI)	Reading (mH$_2$O)	Depth (m)	Reading (PSI)	Reading (mH$_2$O)
1	1.42	1.00	10	14.20	10.00
2	2.84	2.00	15	21.30	15.00
3	4.26	3.00	20	28.54	20.09
4	5.68	4.00	25	35.64	25.09
5	7.10	5.00	30	42.74	30.09

Source: Bo and Choa (2004).

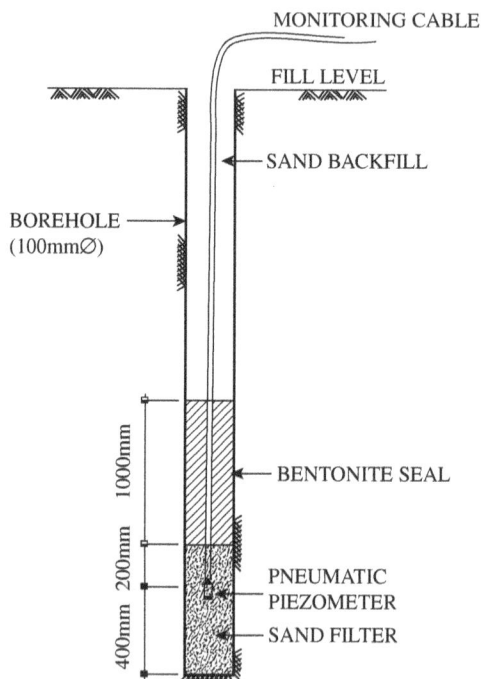

Figure 8.8   Typical installation design of a pneumatic piezometer.
Source: Bo and Choa (2004).

Nowadays, it is increasingly popular to install a string of piezometers in fully grouted holes in a low permeable clay as hydraulic conductivity of the grout mixture is very similar to the surrounding soil. It has been reported by many researchers that the piezometers in the fully grouted boreholes performed in a very similar manner as those installed in the individual boreholes.

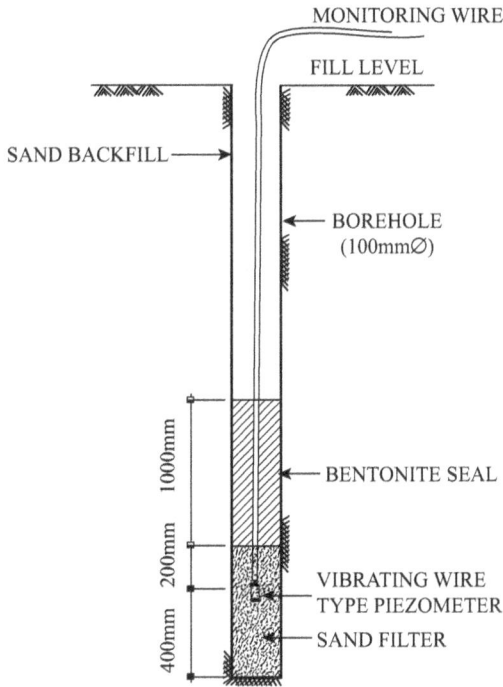

Figure 8.9 Typical installation design of a vibrating wire piezometer.
Source: Bo and Choa (2004).

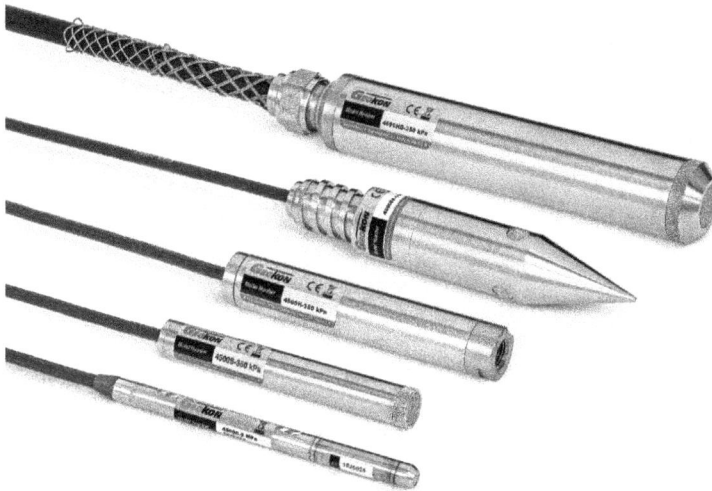

Figure 8.10 Photographic features of pneumatic and vibrating wire piezometers.
Source: Courtesy of GEOKON®, www.geokon.com.

Piezometers generally measure pressure or the water head above the measured level. The measured values are generally translated into a piezometric head or excess pore pressure using static water head. Data are usually presented together with construction stages and activities. Table 8.3 shows measured and processed data, and Figures 8.11 and 8.12 shows data presentation in terms of piezometric elevation, and excess pore pressure.

Basically, piezometers are installed to monitor the dissipation of excess pore pressure.

8.4.2 *Casagrande open type piezometer*

Open type piezometers are installed in a more permeable formation where drainage conditions need to be checked. Open type piezometers are installed in the same manner as pneumatic piezometers. Instead of a pneumatic cable and a water pressure cable, it has an extruding open pipe for water to be floated in the pipe. Water depths are measured with the help of the water level indicator. Sometimes, water could overflow through the pipe due to an

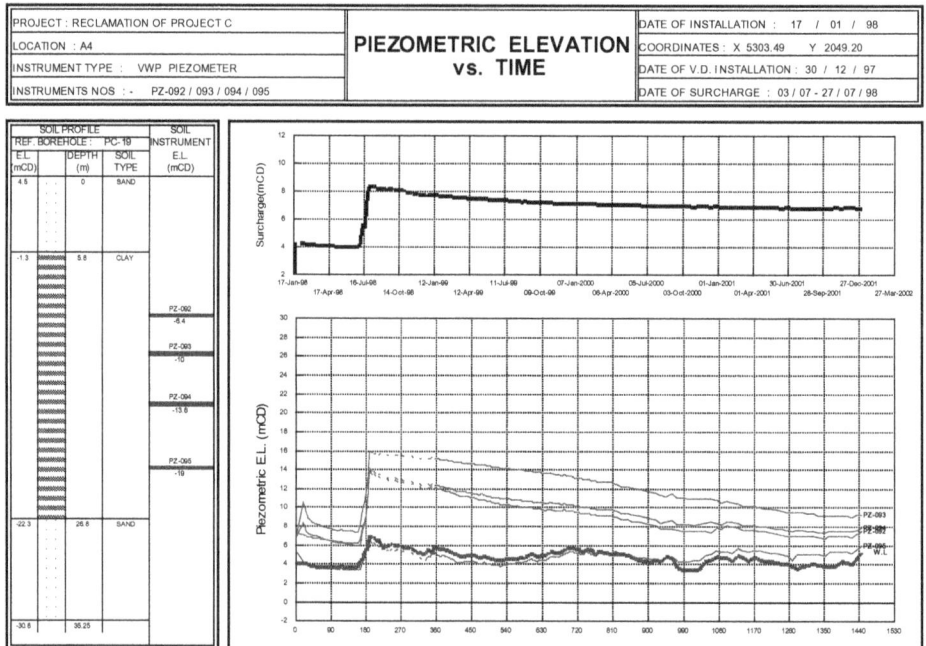

Figure 8.11   Data presentation in terms of piezometer elevation, together with soil profile and construction activities.

Source: Bo and Choa (2004).

Table 8.3   Typical processed data of piezometer monitoring.

EXAMPLE OF VIBRATING WIRE PIEZOMETER MONITORING AND PROCESS DATA

PROJECT: RECLAMATION PROJECT C	DATE OF FILL: 25/11/96-18/01/97
INSTRUMENT NO.: PZ-092	DATE OF V.D. INSTALLATION: 30/12/97
LOCATION: A4 AREA	GROUND E.L.: (+)4.243 mCD
SERIAL NO.: 65072	DATE OF INSTALLATION: 17/01/98
INSTALL E.L. OF PIEZO: -6.40 mCD	COORDINATES: X 5303.49 Y 2049.20
INSTALLATION DEPTH: 10.64 M	DATE OF SURCHARGE: 03/07-27/07/98

No.	DATE	READING TIME	READING INTERVAL (day)	TOTAL TIME (day)	PIEZOMETER READINGS PSI	PIEZOMETER READINGS (kPa)	PIEZO E.L. (mCD)	WSP (WS-A4S06) READING (m)	TOP E.L. (mCD)	WATER LEVEL (mCD)	PORE WATER PRESSURE STATIC (kPa)	EXCESS (kPa)	CORRECTED EXCESS (kPa)	SURCHARGE ELEV. (mCD)	HEIGHT (m)	REMARK
1		16:00	0	4	19.4200	133.901	7.258	0.730	4.733	4.003	101.830	32.071	31.847	4.243	0.00	
2		10:00	14	18	19.3035	133.098	7.176	0.710	4.733	4.023	102.026	31.072	30.848	4.243	0.00	
3		09:40	7	25	19.0028	131.024	6.964	0.720	4.714	3.994	101.742	29.283	28.932	4.248	0.00	
4		09:40	7	32	18.9189	130.446	6.905	0.750	4.701	3.951	101.321	29.125	28.687	4.215	0.00	
5		15:20	7	39	18.8351	129.868	6.8847	0.900	4.685	3.785	99.696	30.172	29.610	4.199	0.00	
6		10:10	7	46	18.7512	129.289	6.788	0.940	4.672	3.732	99.177	30.112	29.463	4.175	0.00	
7		09:00	7	53	18.6673	128.711	6.729	0.950	4.656	3.706	98.923	29.789	29.027	4.160	0.00	
8		09:35	7	60	18.5834	128.133	6.670	0.950	4.640	3.690	98.766	29.367	28.502	4.081	0.00	
9		09:15	7	67	18.4996	127.554	6.611	0.980	4.628	3.648	98.355	29.200	28.263	4.086	0.00	
10		09:07	9	76	18.4157	126.976	6.552	0.920	4.614	3.694	98.805	28.171	27.150	4.115	0.00	
11		08:40	14	90	18.3318	126.398	6.493	0.940	4.588	3.648	98.355	28.043	26.851	4.087	0.00	
12		08:56	7	97	18.2479	125.820	6.434	0.910	4.575	3.665	98.521	27.298	26.027	4.076	0.00	
13		09:20	14	111	18.1641	125.241	6.375	0.850	4.549	3.699	98.854	26.387	24.934	4.028	0.00	
14		08:40	7	118	18.0802	124.663	6.316	0.800	4.543	3.743	99.285	25.378	23.924	4.038	0.00	
15		10:45	7	125	18.0048	124.143	6.263	0.870	4.526	3.656	98.903	25.240	23.718	4.030	0.00	
16		15:30	14	139	17.9671	123.883	6.236	0.870	4.526	3.656	98.433	25.450	23.860	4.042	0.00	
17		16:35	7	146	17.9105	123.493	6.196	0.850	4.503	3.653	98.404	25.089	23.372	4.001	0.00	
18		16:00	7	153	17.8822	123.297	6.176	0.820	4.489	3.669	98.560	24.737	22.917	3.988	0.00	
19		17:55	8	161	17.8538	123.105	6.156	0.830	4.485	3.655	98.423	24.679	22.844	3.986	0.00	
20		16:30	6	167	18.7546	129.251	6.784	2.810	7.242	4.432	106.029	23.222	21.331	3.980	0.00	
21		15:20	3	170	19.6374	135.400	7.411	2.740	7.231	4.491	106.603	28.797	26.875	4.546	0.55	Under surcharge
22		09:18	3	173	20.5292	141.549	8.038	2.750	7.219	4.469	106.394	35.154	33.204	5.112	1.11	

Source: Bo and Choa (2004).

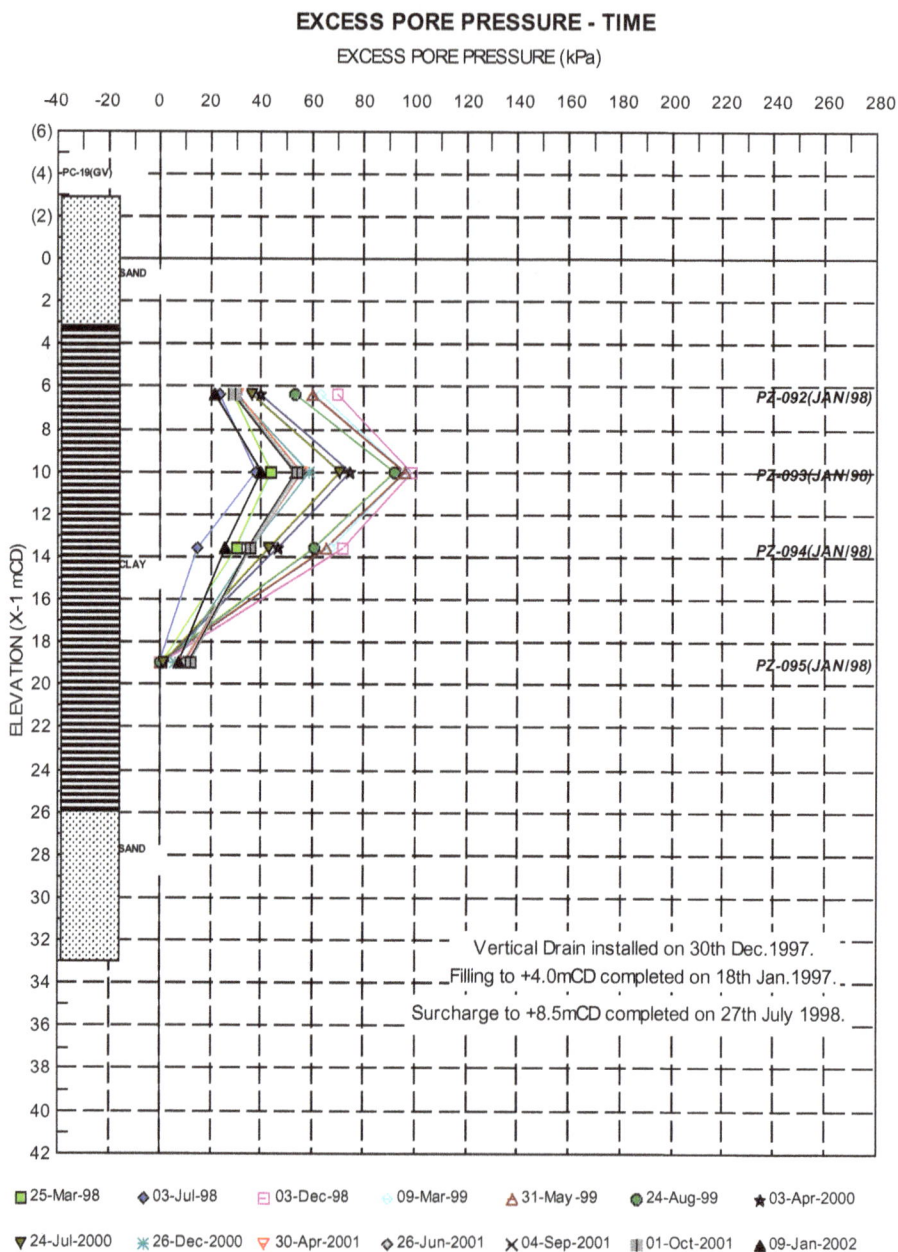

Figure 8.12   Data presentation in terms of excess pore pressures together with soil profile.
Source: Bo and Choa (2004).

extremely high artesian pressure of the aquifer below the compressible layer. As such, a pressure gauge should be installed to measure the water head (Figure 8.13(b)).

A typical installation design is shown in Figure 8.13(a). Figure 8.14 shows a water level indicator used in monitoring of the open type piezometer. Table 8.4 shows measured and processed data. Figure 8.15 shows a graphical presentation of processed data. Data are generally presented in the piezometric elevation. Figure 8.16 shows a photographic feature of a Casagrande open type piezometer.

8.5 Water Standpipe

Water standpipes are installed to measure the static water level of ground water. A water intake open slot would be in the water-bearing formation. Sufficient open area normally greater than 11% should be provided in order to reduce the hydro-dynamic time lag. On the contrary, the opening slot must be small enough to retain the surrounding soil. In normal practice, a geotextile is wrapped around the slotted area in order to be able to retain the surrounding soil. A typical installation design of a water standpipe is shown in Figure 8.17(a). Figure 8.17(b) shows a photographic feature of the water standpipe. The measurements, data processing, and presentation are the same as those with an open type piezometer. Table 8.5 shows

Table 8.4   Measured and processed data obtained from a Casagrande Open Type Piezometer.

CASSAGRANDE OPEN-TYPE PIEZOMETER

PROJECT : RECLAMATION OF PROJECT C
INSTRUMENT NO. : OP-019
LOCATION : A4
INSTALL E.L OF PIEZO : -26.00 mCD
INSTALLATION DEPTH : 30.24 m
GROUND E.L : (+) 4.243 mCD

DATE OF FILL : 05 / 12 / 96 - 18 / 01 / 97
DATE OF V.D. INSTALLATION : 30 / 12 / 97
DATE OF INSTALLATION : 07 / 01 / 98
COORDINATES : X 5301.62   Y 2058.21
DATE OF SURCHARGE : 03 / 07 - 27 / 07 / 98

Sr. No.	DATE	READING TIME	READING INTERVAL (day)	TOTAL TIME (day)	PIEZOMETER READING (m)	PIEZOMETER TOP E.L (mCD)	PIEZO E.L (mCD)	W S P(WS-A4S06) READING (m)	W S P(WS-A4S06) TOP E.L (mCD)	WATER LEVEL (mCD)	EXCESS PORE WATER (P) (kPa)	SURCHARGE ELEV. (mCD)	SURCHARGE HEIGHT (m)	REMARK
1	21-Jan-98	16:00	0	14	0.71	4.816	4.106	0.73	4.733	4.003	1.03	4.243	0	
2	4-Feb-98	10:00	14	28	0.76	4.816	4.056	0.71	4.733	4.023	0.33	4.243	0	
3	11-Feb-98	9:40	7	35	0.76	4.81	4.05	0.72	4.714	3.994	0.56	4.248	0	
4	18-Feb-98	9:40	7	42	0.82	4.809	3.989	0.75	4.701	3.951	0.38	4.215	0	
5	25-Feb-98	15:20	7	49	0.98	4.804	3.824	0.9	4.685	3.785	0.39	4.199	0	
6	4-Mar-98	10:10	7	56	1.04	4.801	3.761	0.94	4.672	3.732	0.29	4.175	0	
7	11-Mar-98	9:00	7	63	1.04	4.796	3.756	0.95	4.656	3.706	0.5	4.16	0	
8	18-Mar-98	9:35	7	70	1.08	4.79	3.71	0.95	4.64	3.69	0.2	4.081	0	
9	25-Mar-98	9:15	7	77	1.05	4.791	3.741	0.98	4.628	3.648	0.93	4.086	0	
10	3-Apr-98	9:07	9	86	1.07	4.787	3.717	0.92	4.614	3.694	0.23	4.115	0	
11	17-Apr-98	8:40	14	100	1.1	4.779	3.679	0.94	4.588	3.648	0.31	4.087	0	
12	24-Apr-98	8:56	7	107	1.08	4.775	3.695	0.91	4.575	3.665	0.3	4.076	0	
13	8-May-98	9:20	14	121	1.05	4.762	3.712	0.85	4.549	3.699	0.13	4.028	0	
14	15-May-98	8:40	7	128	0.98	4.763	3.783	0.8	4.543	3.743	0.4	4.038	0	
15	22-May-98	10:45	7	135	0.81	4.761	3.951	0.83	4.534	3.704	2.47	4.03	0	
16	5-Jun-98	15:30	14	149	1.02	4.761	3.741	0.87	4.526	3.656	0.85	4.042	0	
17	12-Jun-98	16:38	7	156	1.01	4.754	3.744	0.85	4.503	3.653	0.91	4.001	0	
18	19-Jun-98	16:00	7	163	0.96	4.739	3.779	0.82	4.489	3.669	1.1	3.986	0	
19	27-Jun-98	17:55	8	171	0.98	4.741	3.761	0.83	4.485	3.655	1.06	3.986	0	
20	3-Jul-98	16:30	6	177	2.83	7.514	4.684	2.81	7.242	4.432	2.52	3.96	0	
21	8-Jul-98	15:20	3	180	2.75	7.516	4.766	2.74	7.231	4.491	2.75	4.546	0.55	Under surcharge
22	9-Jul-98	9:18	3	183	2.77	7.518	4.748	2.75	7.219	4.469	2.79	5.112	1.11	

Source: Bo and Choa (2004).

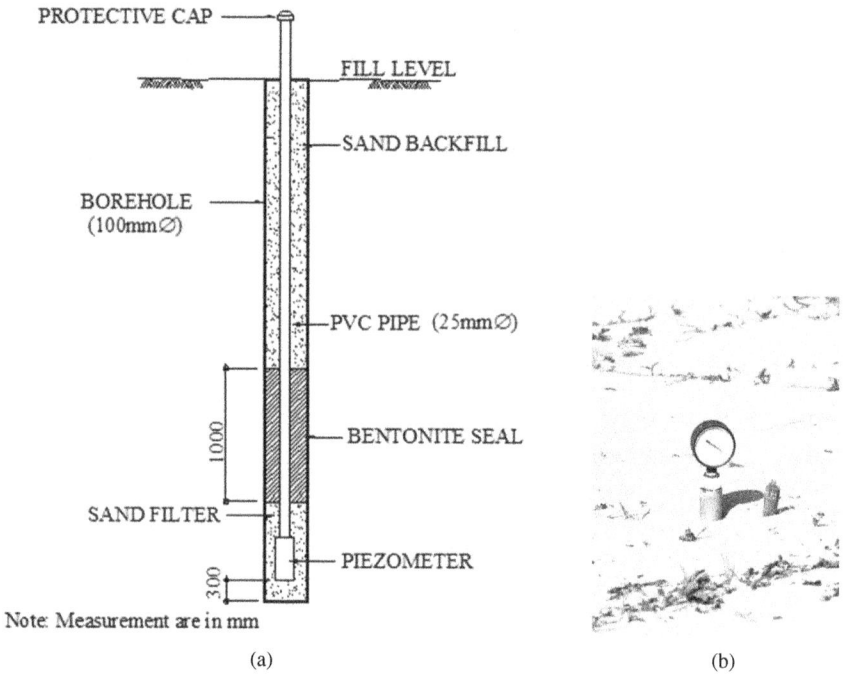

(a)   (b)

Figure 8.13   (a) Typical installation of an open type piezometer, (b) Casagrande open type piezometer with a pressure gauge.

Source: Bo and Choa (2004).

Figure 8.14   Water level indicator used in monitoring of a water standpipe.

Source: Courtesy of GEOKON®, www.geokon.com.

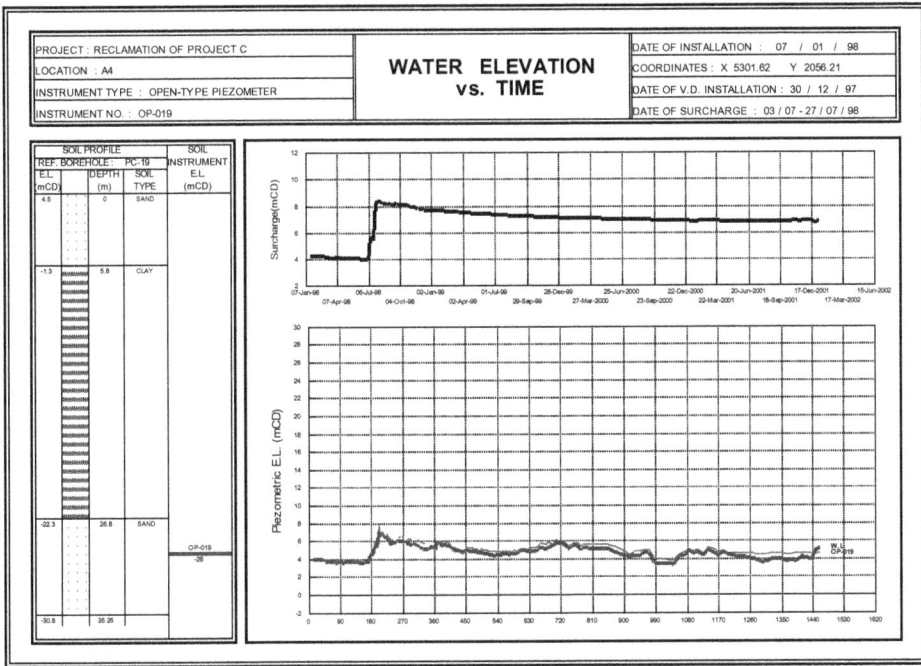

Figure 8.15   Graphical presentation of processed water level data.

Source: Bo and Choa (2004).

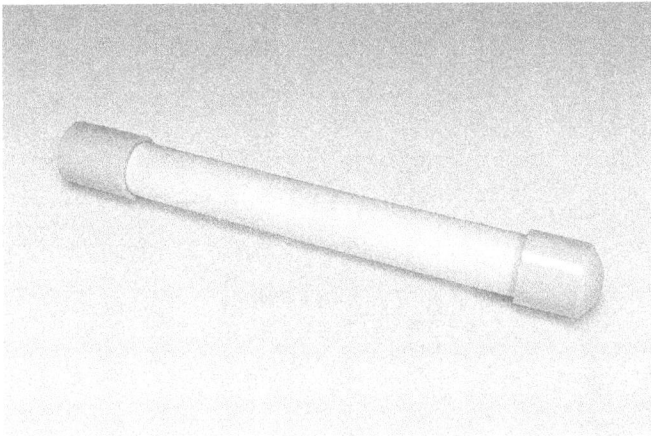

Figure 8.16   Photographic feature of a Casagrande open type piezometer.

Source: Bo and Choa (2004) (courtesy of GEOKON® | www.geokon.com).

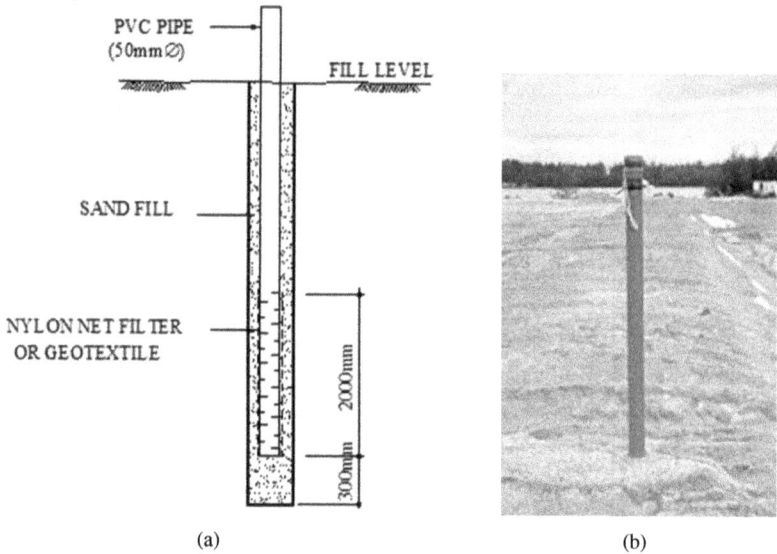

Figure 8.17   (a) Typical installation design feature and (b) photographic feature.
Source: Bo and Choa (2004).

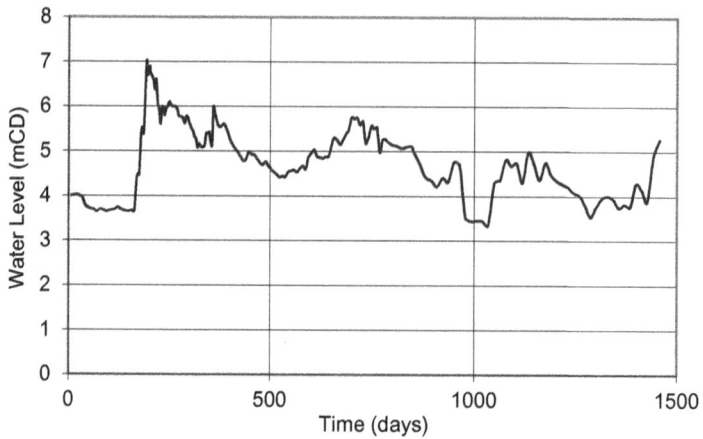

Figure 8.18   Example of data presentation.
Source: Bo and Choa (2004).

processed data and Figure 8.18 shows an example of the data presentation. In many cases where groundwater sampling is not required, a 25-mm-diameter water standpipe is installed. Where groundwater sampling is required for water chemistry analyses for the purposes of a discharge permit, a minimum 50-mm-diameter water standpipe is installed.

Table 8.5   Water standpipe processed monitoring data.

					W S P(WS-A4S06)		PORE WATER(P)		SURCHARGE	
WATER STAND PIPE RECORD SHEET										
PROJECT : RECLAMATION AT PROJECT A						DATE OF FILL : 05 / 12 / 96 - 18 / 01 / 97				
LOCATION : A4S-06 AREA						DATE OF V.D. INSTALLATION : 30 / 12 / 97				
						GROUND E.L : (+) 4.243 mCD				
						DATE OF SURCHARGE : 03 / 07 - 27 / 07 / 98				
No.	DATE	READING TIME	READING INTERVAL (day)	TOTAL TIME (day)	TOP E.L (mCD)	WATER LEVEL (mCD)	STATIC (kPa)	EXCESS (kPa)	ELEV. (mCD)	HEIGHT (m)
1	21-Jan-98	16:00	0	7	4.733	4.003	101.83	49.86	4.243	0
1	21-Jan-98	16:00	0	7	4.733	4.003	101.83	49.86	0	
1	21-Jan-98	16:00	0	7	4.733	4.003	101.83	49.86		
2	4-Feb-98	10:00	14	21	4.733	4.023	102.026	46.217	4.243	0
3	11-Feb-98	9:40	7	28	4.714	3.994	101.742	47.19	4.248	0
4	18-Feb-98	9:40	7	35	4.701	3.951	101.321	45.543	4.215	0
5	25-Feb-98	15:20	7	42	4.685	3.785	99.696	32.688	4.199	0
6	4-Mar-98	10:10	7	49	4.672	3.732	99.177	42.86	4.175	0
7	11-Mar-98	9:00	7	56	4.656	3.706	98.923	43.804	4.16	0
8	18-Mar-98	9:35	7	63	4.64	3.69	98.766	51.545	4.081	0
9	25-Mar-98	9:15	7	70	4.628	3.648	98.355	44.372	4.086	0
10	3-Apr-98	9:07	9	79	4.614	3.694	98.805	43.232	4.115	0
11	17-Apr-98	8:40	14	93	4.588	3.648	98.355	44.372	4.087	0
12	24-Apr-98	8:56	7	100	4.575	3.665	98.521	43.516	4.076	0
13	8-May-98	9:20	14	114	4.549	3.699	98.854	43.872	4.028	0
14	15-May-98	8:40	7	121	4.543	3.743	99.285	43.442	4.038	0
15	22-May-98	10:45	7	128	4.534	3.704	98.903	41.755	4.03	0
16	5-Jun-98	15:30	14	142	4.526	3.656	98.433	39.467	4.042	0
17	12-Jun-98	16:35	7	149	4.503	3.653	98.404	42.254	4.001	0
18	19-Jun-98	16:00	7	156	4.489	3.669	98.56	42.098	3.988	0
19	27-Jun-98	17:55	8	164	4.485	3.655	98.423	43.614	3.986	0
20	3-Jul-98	16:30	6	170	7.242	4.432	106.029	38.076	3.98	0
21	6-Jul-98	15:20	3	173	7.231	4.491	106.603	41.639	4.546	0.55
22	9-Jul-98	9:18	3	176	7.219	4.469	106.394	42.538	5.112	1.11
23	13-Jul-98	15:30	4	180	7.208	5.328	114.8	56.886	5.678	1.68
24	16-Jul-98	8:00	3	183	7.213	5.523	116.708	54.977	5.55	1.55
25	20-Jul-98	15:20	4	187	9.34	5.39	115.406	56.279	5.421	1.42
26	23-Jul-98	14:30	3	190	9.349	6.049	121.857	55.344	7.061	3.06
27	27-Jul-98	16:00	4	194	9.294	7.014	131.303	91.406	8.335	4.34
28	30-Jul-98	8:15	3	197	9.277	6.687	128.102	92.538	8.343	4.34
29	4-Aug-98	14:38	5	202	9.26	6.89	130.089	93.309	8.35	4.35
30	6-Aug-98	14:25	2	204	9.246	6.746	128.675	90.586	8.348	4.35

Source: Bo and Choa (2004).

8.6 Inclinometer

Inclinometers are installed to monitor the lateral movement of the ground at the edge of the slope or retaining structure. Generally, an inclinometer is installed at the trough of the potential slip circle line within the zone of expected soil mass movement. The inclinometer should also be installed on the crest and the toe of the embankment. A typical arrangement of inclinometer installation at the slope and embankment is shown in Figure 8.19 and one behind the retaining structure is shown in Figure 8.20. Settlement plates are also usually installed together in order to be able to relate lateral displacement with vertical displacement. For the vertical wall, inclinometers are attached to or installed behind the inner side of the wall.

An inclinometer should be anchored at hard competent formation where there is no lateral movement. Since an inclinometer measures relative movement to the toe, any movement at the toe would lead to the underestimation of lateral displacement.

A typical installation design of an inclinometer is shown in Figure 8.21.

The inclinometer monitoring probe generally measures inclination degree between two points, and lateral displacement is calculated using the following simple equation:

$$Horizontal\ Displacement = L \times \sin \theta \tag{8.1}$$

where L is the length measured along the casing,
          θ is the angle from the vertical line.

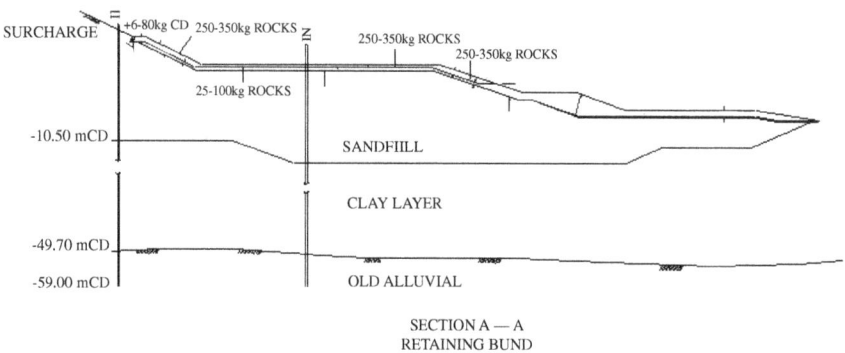

Figure 8.19   Typical arrangement of inclinometer installation at a shore protection structure.
Source: Bo and Choa (2004).

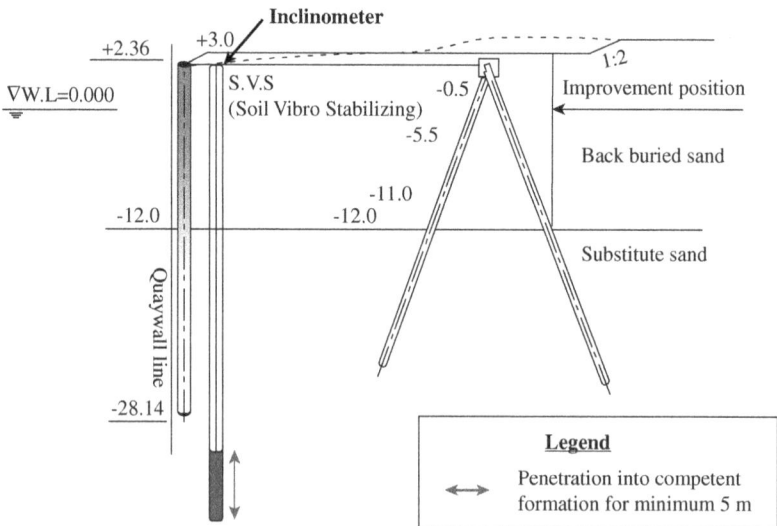

Figure 8.20   Typical Inclinometer installation behind a retaining structure.

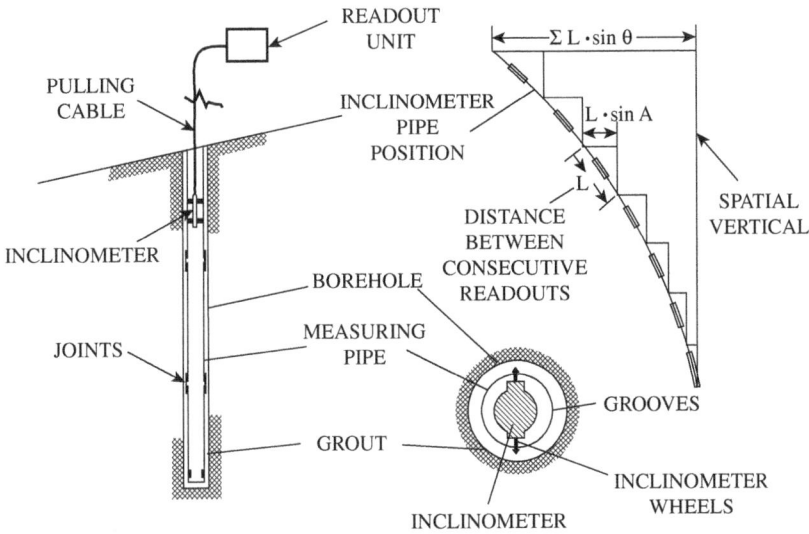

Figure 8.21   Typical design and principal of an inclinometer.
Source: Bo and Choa (2004) Courtesy of Slope Indicator.

Cumulative displacements are calculated from the measurement. Table 8.6 shows an example of processed data and Figure 8.22 shows an example of data presentation. Figure 8.23 shows photographic features of various types of coupling and casing used in inclinometer installation. Figure 8.24 shows an inclinometer probe used for monitoring and Figure 8.25 shows an inclinometer spiral sensor for checking the casing twist. Figure 8.26 shows construction control using inclinometer monitoring data such as magnitude of displacement and rate of displacement.

8.7 Gas Monitoring Well

If ground investigation is carried out on the contaminated land, gas monitoring may be required. A gas monitoring well is very similar to a water standpipe. Through the gas monitoring well, gas sampling can be carried out for further analyses. A guide on the number of gas monitoring wells required and their spacings was described by Wilson *et al.* (2009) and is shown in Table 8.7. A typical gas monitoring well is shown in Figure 8.27.

Table 8.6   Processed data from inclinometer measurements.

DIRECTION DEPTH (mCD) (70.0 m)	SET40 22-Sep-99 22/09:1416~1435 Disp	SET70 22-Jan-00 22/01:1439~1509 Disp	SET89 29-Apr-00 29/04:1559~1620 Disp	SET106 24-Aug-00 24/08:1639~1701 Disp	SET142 18-May-01 18/05:1003~1030 Disp	SET152 27-Jul-01 27/07:0854~0911 Disp	SET158 12-Oct-01 2/10:0950~1015 Disp
6.03	-38.97	-114.24	-138.08	-112.25	-142.96	-146.59	-146.32
5.03	-37.52	-113.64	-138.08	-109.44	-142.72	-145.56	-145.76
4.03	-39.66	-107.98	-118.84	-88.36	-117.52	-113.88	-112.36
3.03	-33.38	-91.48	-98.38	-63.2	-85.98	-81.64	-79.42
2.03	-28.8	-78.22	-80.72	-40.86	-57.54	-52.62	-49.48
1.03	-24.04	-66.08	-64.02	-19.48	-30.42	-24.52	-20.54
0.03	-18.98	-53.58	-46.88	2.22	-3.2	3.66	8.42
-0.97	-13.76	-40.48	-29.22	24.42	24.44	32.22	37.78
-1.97	-8.56	-27.36	-11.5	46.72	52.42	60.74	67.1
-2.97	-3.6	-14.6	5.8	68.56	79.5	88.82	95.98
-3.97	0.1	-3.8	21.26	88.26	104	114.54	122.62
-4.97	3.88	8.04	37.86	109.06	130.42	141.62	150.64
-5.97	10.02	22.66	57.68	133	160.58	171.9	181.64
-6.97	15.1	35.74	75.64	155.26	188.48	200.5	211.02
-7.97	19.24	47.22	92.04	176.04	214.34	227.62	239.06
-8.97	25.18	61.42	111.08	199.62	243.48	257.58	269.76
-9.97	28.66	72.6	127.08	219.94	269.96	284.48	297.58
-10.97	33.06	84.54	143.76	241.06	296.8	311.98	325.9
-11.97	38.4	98.38	162.62	264.48	325.88	342.02	356.82
-12.97	43.14	110.52	179.42	285.76	352.16	369.1	384.7
-13.97	47.9	124	198.02	308.98	381.42	399.12	415.52
-14.97	52.22	136.86	215.7	331.44	410.26	428.5	445.42
-15.97	57.12	149.26	232.66	352.08	435.56	454.08	471.6
-16.97	62.48	162.24	250.06	373.16	462.18	481.12	499.26
-17.97	64.44	167.9	258.08	382.48	473.96	493.12	511.86
-18.97	67.08	171.44	262.36	387.98	479.9	499.86	518.98
-19.97	66.56	170.02	260.78	385.48	476.56	496.48	515.96
-20.97	66.72	170.18	260.68	382.76	473.62	492.82	502.98
-21.97	64.36	167.64	254.78	371.66	461.36	478.12	489.1
-22.97	64.12	166.36	252.08	365.74	454.04	469.62	480.32
-23.97	62.42	162.16	246.78	360.52	446.8	464.14	474.6
-24.97	58.68	155.5	237.12	346.48	430.6	446.82	456.96
-25.97	58.68	153.72	233.6	339.26	422.02	435.98	438.72
-26.97	54.66	145.64	221.4	323.46	403.58	417.68	422.92
-27.97	50.86	137.5	209.48	307.92	386.7	402.1	410.58
-28.97	47.38	130.1	199.06	294.14	369.96	384.3	392.22
-29.97	42.28	118.92	181.8	271	343.6	357.7	366.42
-30.97	36.18	105.32	163.2	247.38	315.8	329.24	337.52
-31.97	32.56	95.3	145.8	218.14	275.64	285.88	292.1
-32.97	30.26	88.06	135.32	202.74	256.4	265.76	270.52
-33.97	31.96	85.94	126	178.28	214.66	220.22	222.6
-34.97	29.4	77.98	112.34	151.66	175.1	177.84	176.44
-35.97	22.88	60.92	86.22	114.18	134.04	136.8	136.4
-36.97	17.1	42.24	57.38	76.68	93.3	96.74	99.02
-37.97	12.94	28.46	34.26	45.88	60.32	64.44	67.88
-38.97	9.54	18.7	22.08	32.72	47.34	51.86	56.34
-39.97	7.3	14.76	18.08	27.54	39.34	43.72	49.22
-40.97	7.26	13.34	15.74	22.72	29.52	33.14	38.28
-41.97	4.74	12.92	18.14	27.66	35.2	38.54	42.74
-42.97	3.6	10.26	14	21.44	29.02	32.48	36.64
-43.97	4.02	10.78	14.34	21.62	29.74	33.26	37.62
-44.97	4.08	11.66	16	23.38	30.94	34.2	37.3
-45.97	3.14	7.42	9.26	13.3	20.7	24.88	29.76
-46.97	1.16	7.2	9.14	11.58	13.96	17.04	20.62
-47.97	0.62	3.12	3.86	7.96	14.56	20.04	27.86
-48.97	4.16	10.94	14.26	20.64	28.64	33.62	39.28
-49.97	3.94	11.14	14.26	19.24	23.34	27.02	30.08
-50.97	2.22	7.42	9.74	16.42	23.18	27.66	31.44
-51.97	3.86	8.64	8.34	8.04	-4.88	-7.36	-10.6
-52.97	4.14	3.88	2.5	7.16	8.1	8.86	14.12
-53.97	3.12	5.96	6.8	11.52	14.78	16.24	16.54
-54.97	1.76	4.02	2.64	1.76	-0.2	0.12	-1.4
-55.97	-0.94	-1.3	-4.58	-6.64	-8.38	-7.88	-8.9
-56.97	-2.4	-1.86	-4.66	-7.3	-9.78	-9.5	-10.6
-57.97	-2.06	-2.96	-5.42	-5.98	-7.94	-7.56	-7.8
-58.97	-1.98	-2.24	-3.56	-2.34	-2.68	-2.16	-1.82
-59.97	-0.24	0.78	0.18	2.62	3.46	3.6	4.54
-60.97	0.56	2.26	2.04	3.92	5.5	6.06	7.3
-61.97	1.12	2.74	3.1	5.5	7.6	8.16	9.02

Source: Bo and Choa (2004).

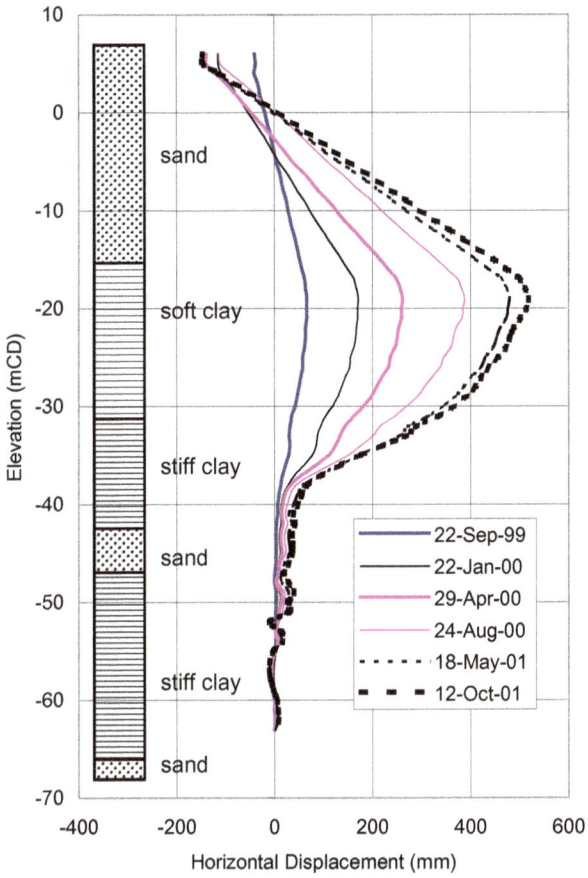

Figure 8.22   Example data presentation for inclinometer measurements.
Source: Bo and Choa (2004).

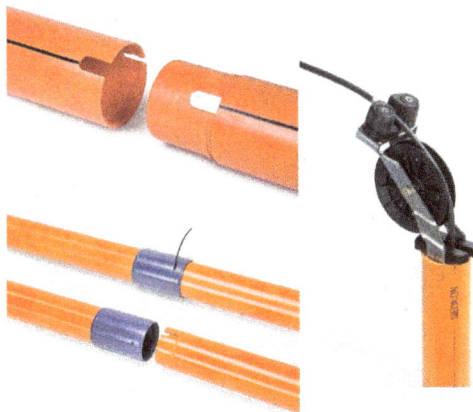

Figure 8.23   Photographic features of various types of coupling/and casing.
Source: Courtesy of GEOKON®, www.geokon.com.

Figure 8.24   Inclinometer probe used for monitoring.
Source: Courtesy of GEOKON®, www.geokon.com.

Figure 8.25   Using the spiral sensor for checking the casing twist.
Source: Bo and Choa (2004).

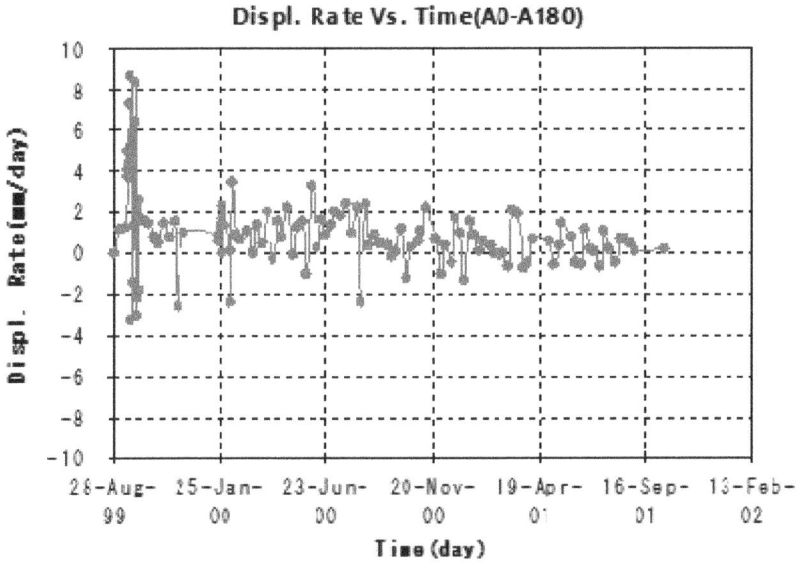

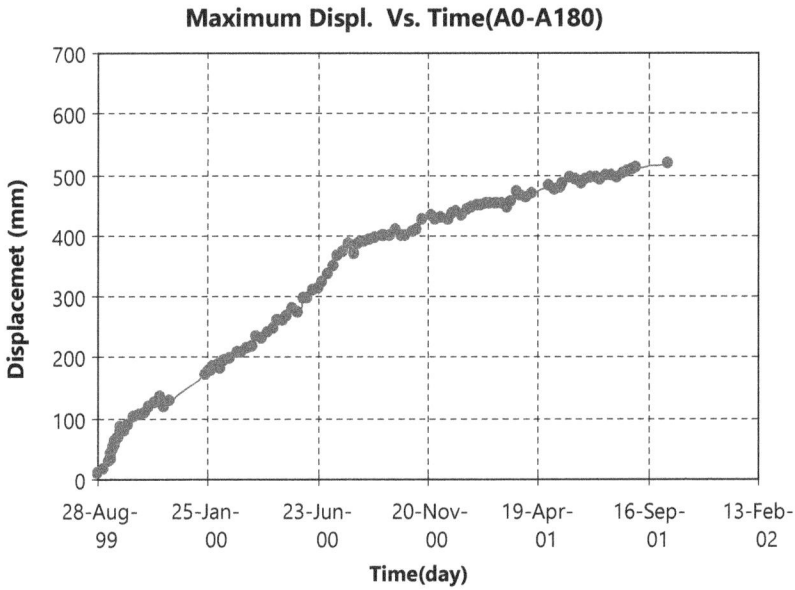

Figure 8.26   Construction control using inclinometer monitoring data.

Source: Bo and Choa (2004).

Table 8.7　Monitoring well spacing to detect off-site gas migration.

Site Description	Monitoring Borehole Spacing (m) Typical Range
Uniform low permeability strata (for example, clay); no development within 250 m	50–150
Uniform low permeability strata (for example, clay); development within 250 m	20–50
Uniform low permeability strata (for example, clay); development within 150 m	10–50
Uniform matrix dominated permeable strata (for example, porous sandstone); no development within 250 m	20–50
Uniform matrix dominated permeable strata (for example, porous sandstone); development within 250 m	10–50
Uniform matrix dominated permeable strata (for example, porous sandstone); development within 150 m	10–20
Fissure or fracture flow dominated permeable strata (for example, blocky sandstone or igneous rock); no development within 250 m	20–50
Fissure or fracture flow dominated permeable strata (for example, blocky sandstone or igneous rock); development within 250 m	10–50
Fissure or fracture flow dominated permeable strata (for example, blocky sandstone or igneous rock); development within 150 m	5–20

Source: Wilson *et al.* (2009).

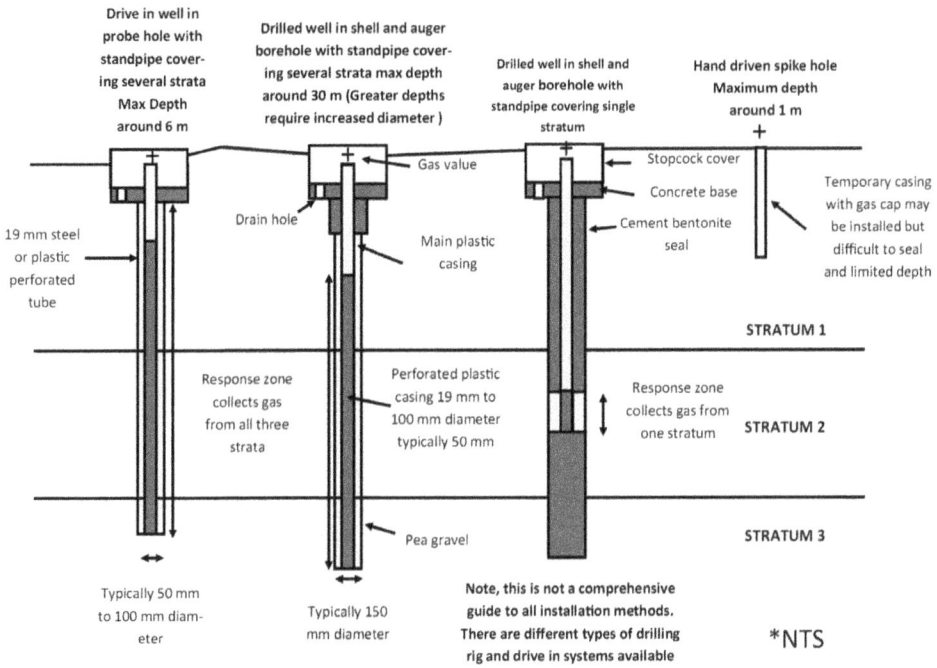

Figure 8.27   Typical gas well installations.
Source: Wilson *et al.* (2009).

Chapter 9

Data Management and Reporting

After carrying out the ground investigation and after the necessary information and data are collected, the next step will be presenting and interpreting those data as factual information and the interpreted results in a form of a graphical presentation and report. These can be used for decision making and input for geotechnical design.

9.1 Field Description

Field descriptions based on the collected samples are usually conducted by drillers or supervising engineering geologists based on the visual and physical inspections in the field and immediately available *in-situ* test results. These descriptions have been termed as driller's log in the past. The details of field description of engineering soils and rocks have been extensively described by West (1990). Field estimation of compactness and strength of soils has been described in BS 5930:1981 and is shown in Table 9.1. Field description of soil is normally started by differentiation between granular and cohesive soils by visualization and feeling and crumbling by hand to estimate the grain size and cohesiveness. Their consistencies such as compactness and strength can be determined from field *in-situ* tests such as the Standard Penetration Test and Field Vane Shear Test. In some cases where lumps of relatively undisturbed cohesive soil are available, handheld vane equipment (Figure 9.1) and handheld penetrometers (Figure 9.2) can be used to estimate the strength of the cohesive soils. Tables 9.2 and 9.3 show field identification of basic soil types described for granular soils and cohesive soils in BS 5930: 1981, respectively. Field identification applying the Unified Soil Classification system is shown in Table 9.4.

Table 9.1   Field estimation of compactness and strength of soils.

		Term	Field Test
Granular Soils	Boulders and Cobbles	Loose	By inspection of voids and particle packing.
		Dense	
	Gravels and Sands	Loose	Can be excavated with a spade; 50-mm wooden peg can be easily driven.
		Dense	Requires pick for excavation; 50-mm wooden peg hard to drive.
		Slightly cemented	Visual examination; pick removes soil in lumps which can be abraded.
	Silts	Soft or loose	Easily molded or crushed in the fingers.
		Firm or dense	Can be molded or crushed by strong pressure in the fingers.
Cohesive Soils	Clays	Very soft	Exudes between fingers when squeezed in hand.
		Soft	Molded by light finger pressure.
		Firm	Can be molded by strong finger pressure.
		Stiff	Cannot be molded by fingers. Can be indented by thumb.
		Very stiff	Can be indented by thumb nail.
	Peat	Firm	Fibers already compressed together.
		Spongy	Very compressible and open structure.
		Plastic	Can be molded in hand, and smears fingers.

Source: Adapted from BS 5930:1981 by West 1990.

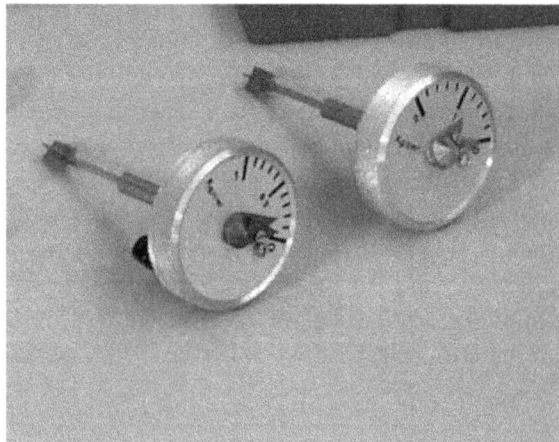

Figure 9.1   Various types of handheld vane equipment.
Source: Courtesy of Matest S.p.A.

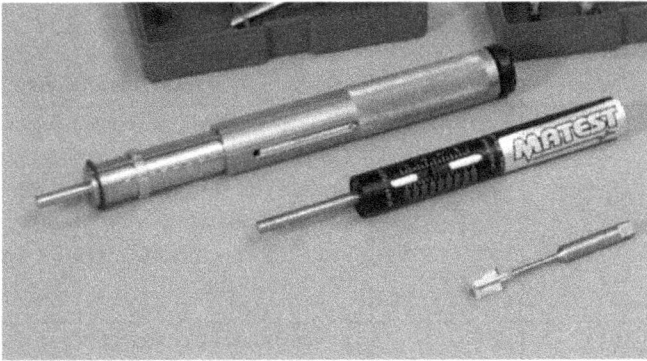

Figure 9.2　Various types of handheld penetrometers.
Source: Courtesy of Matest S.p.A.

Table 9.2　Field identification of basic soil types, granular soils.

		Basic Soil Types	Particle Size (mm)	Visual Identification
Granular Soils	Very Coarse Soils	Boulders	>200	Only seen complete inputs or exposures.
		Cobbles	200 > 60	Often difficult to recover from boreholes.
	Coarse Soils (over 65% sand and gravel sizes)	Gravel	Coarse 60 > 20	Easily visible to naked eye; particle shape can be described; grading can be described.
			Medium 20 > 6	Well graded: wide range of grain sizes, well distributed. Poorly graded: not well graded.
			Fine 6 > 2	(May be uniform: size of most particles lies between narrow limits; or gap graded: an intermediate size of particle is missing.)
		Sand	Coarse 2 > 0.6	Visible to naked eye; very little or no cohesion when dry; grading can be described.
			Medium 0.6 > 0.2	Designated well or poorly graded as for gravel.
			Fine 0.2 > 0.06	

Source: Adapted from BS 5930:1981 by West 1990.

Tables 9.5 and 9.6 show the determination of consistencies of soil applying SPT and FVT measurements. The color of the soil can be identified visually or more accurately by comparison with a soil color chart, which is commercially available (Figure 9.3).

Table 9.3   Field identification of basic soil types, cohesive soils.

		Basic Soil Types	Particle Size (mm)	Visual Identification
Cohesive Soils	Fine Soils (over 35 % silt and clay sizes)	Silt	Coarse 0.06 > 0.02	Only coarse silt barely visible to naked eye; exhibits little plasticity and marked dilatancy; slightly granular or silky to the touch. Disintegrates in water; lumps dry quickly; possess cohesion but can be powdered easily between fingers.
			Medium 0.02 > 0.006	
			Fine 0.006 > 0.002	
		Clay	<0.002	Dry lumps can be broken but not powdered between the fingers; they can also disintegrate under water but more slowly than silt; smooth to the touch; exhibits plasticity but no dilatancy; sticks to the fingers and dries slowly; shrinks appreciably on drying usually showing cracks.
	Organic soils	Organic Clay, Silt, or Sand	Varies	Contains substantial amounts of organic vegetable matter.
		Peat	Varies	Predominately plant remains, usually dark brown or black in color, often distinctive smell; low bulk density.

Source: Adapted from BS 5930:1981 by West 1990.

Table 9.4   Field identification of soils of the Unified Soil Classification system (ASTM–D2487).

		Group Symbols	Field Identification Criteria
Granular Soils	Gravels	(GW)	If the gravel is clean, decide if it is well graded (W) or poorly graded (P) by estimating particle size present.
		(GP)	
		(GM)	If the gravel contains fines, decide if the fines are silty (M) or clayey (C) by feel.
		(GC)	
	Sands	(SW)	Individual grains of sand can be distinguished by the naked eye. If the sand is clean, proceed as above (W or P). If the sand contains fines, proceed as above (M or C).
		(SP)	
		(SM)	
		(SC)	
Cohesive Soils	Silts and Clays	(ML)	If soil is of low plasticity, decide if it is silt (M), clay (C), or organic (O) on the basis of dilatancy (silt) and color (organic soils will be very dark).
		(CL)	
		(OL)	
		(MH)	If the soil is of high plasticity, decide if it is silt (M), clay (C), or organic (O) as above.
		(CH)	
		(OH)	
		(PT)	Contains recognizable fibrous material. Dark.

Table 9.5 Consistency and undrained shear strength of cohesive soils.

Consistency	Undrained Shear Strength (kPa)	SPT N-Index (blows/0.3 m)
Very soft	<12	<2
Soft	12–25	2–4
Firm	25–50	4–8
Stiff	50–100	8–15
Very stiff	100–200	15–30
Hard	>200	>30

Source: Canadian Foundation Manual (2006).

Table 9.6 Compactness condition of sands from standard penetration test.

Compactness Condition	SPT N-INDEX (blow/0.3 m)
Very loose	0–4
Loose	4–10
Compact	10–30
Dense	30–50
Very dense	Over 50

Source: Canadian Foundation Manual (2006).

9.2 Driller's Logs vs. Engineer's Logs

Driller's logs are upgraded to Engineer's logs after obtaining classification test results such as moisture content (BS 1377 Part 2 1990, ASTM D 2216-19), Atterberg limits (BS 1377 Part 2 1990, ASTM D 4318-17e1), grain size distribution (BS 1377 Part 2 1990 , ASTM D 6913M-17 and ASTM D7918-17), and organic content (BS 1377 Part 3 1990, ASTM D 2974-20e1). After obtaining those laboratory results, soil classification can be updated to the Engineer's log applying widely acceptable standards such as the British Soil Classification System (BSCS) or Unified Soil Classification System (USCS) and Euro Code Standard as shown in Tables 9.7–9.9, respectively. Details of the USCS system can be found in ASTM D 2487-17e1.

Brief descriptions of soil and its respective order in presentation of borehole log in various countries are shown in Table 9.10.

A complete borehole log should include following information:

1. Project Name
2. Project Coordinate

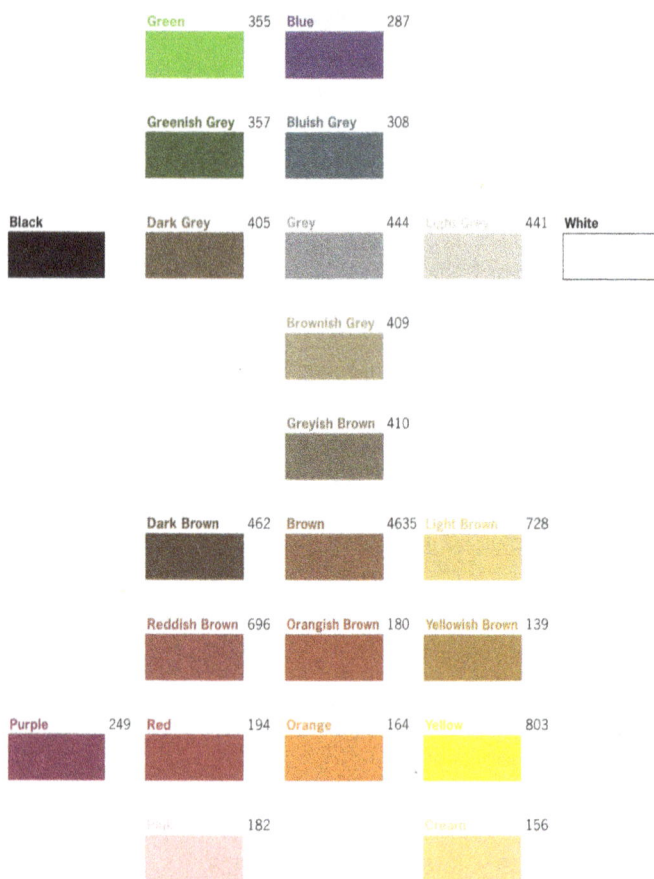

| Green | 355 | Blue | 287 |
| Greenish Grey | 357 | Bluish Grey | 308 |

Black

| Dark Grey | 405 | Grey | 444 | Light Grey | 441 | White |

Brownish Grey 409

Greyish Brown 410

Dark Brown	462	Brown	4635	Light Brown	728
Reddish Brown	696	Orangish Brown	180	Yellowish Brown	139
Purple	249	Red	194	Orange	164	Yellow	803
Pink	182			Cream	156

Figure 9.3   Soil color chart.
Source: Norbury (2010).

3. Type of Drill Rig and method of drilling used
4. Diameter of drill hole
5. Groundwater level measurements
6. Any geotechnical instrument such as water standpipe installation
7. Depths to the top of formations
8. Elevation of formations
9. *In-situ* measurement details such as SPT, FVT
10. Moisture content
11. Atterberg limits test results
12. Grain size distribution percentage
13. Special note on drilling responses

A typical borehole log is shown in Figure 9.4.

Table 9.7 The Unified Soil Classification System.

Major Divisions			Group Symbol	Group Name
Coarse-grained soils; more than 50% retained on 0.0075 mm sieve	*Gravel* > 50% of coarse fraction retained on 4.75 mm sieve	Clean gravel < 5% smaller than 0.075 mm sieve	GW	Well-graded gravel, fine to coarse gravel
			GP	Poorly graded gravel
		Gravel with > 12% fines	GM	Silty gravel
			GC	Clayey gravel
	Sand >= 50% of coarse fraction passes 4.75 mm sieve	Clean sand	SW	Well-graded sand, fine to coarse sand
			SP	Poorly graded sand
		Sand with > 12% fines	SM	Silty sand
			SC	Clayey sand
Fine-grained soils; more than 50% passes 0.075 mm sieve	Silt and clay liquid limit < 50	Inorganic	ML	Silt
			CL	Clay
		Organic	OL	Organic silt, organic clay
	Silt and clay liquid limit >= 50	Inorganic	MH	Silt with high plasticity, elastic silt
			CH	Clay with high plasticity, fat clay
		Organic	OH	Organic clay, organic silt
Organic soils			Pt	Peat

Key to abbreviations

G gravel
S sand
M silt
C clay
O organic

P poorly graded (uniform particle sizes)
W well graded (range of particle sizes)
H high plasticity
L low plasticity

Source: Norbury (2010).

9.3 Cross Section

In order to visualize a two-dimensional profile view of sub-surface succession along the selected sections, cross sections are usually prepared using engineer's borehole logs available along the selected sections. Figure 9.5 show an example of a soil profile cross section along selected sections using borehole data. The cross section of the soil profile generally consists of a profile for each individual formation together with a topographic profile of the existing ground surface and a profile

Table 9.8   The British standard soil classification system.

Soils Groups		Description and Identification		Sub-Groups		Liquid Limit	Fines (% < 0.06 mm)
Slightly silty or clayey Gravel	G	Well-graded Gravel	GW	—			0–5
		Poorly graded Gravel	GP	Uniformly graded	GPu		
				Gap graded	GPg		
Silty or clayey Gravel	G-F	Silty Gravel	G-M	Well graded	GWM		5–15
				Poorly graded	GPM		
		Clayey Gravel	G-C	Well graded	GWC		
				Poorly graded	GPC		
Very silty or clayey Gravel	GF	Very silty Gravel	GM	Subdivisions as for GC	GML, etc.	As GC	15–35
		Very clayey Gravel	GC	Low plasticity fines	GCL	<35	
				Intermediate plasticity fines	GCI	35–50	
				High plasticity fines	GCH	50–70	
				Very high plasticity fines	GCV	70–90	
				Extremely high plasticity fines	GCE	>90	
Slightly silty or clayey Sand	S	Well-graded Sand	SW	—			0–5
		Poorly graded Sand	SP	Uniformly graded	SPu		
				Gap graded	SPg		
Silty or clayey Sand	S-F	Silty Sand	S-M	Well graded	SWM		5–15
				Poorly graded	SPM		
		Clayey Sand	S-C	Well graded	SWC		
				Poorly graded	SPC		
Very silty or clayey Sand	SF	Very silty Sand	SM	Subdivisions as for SC	SML, etc.	As SC	15–35
		Very clayey Sand	SC	Low plasticity fines	SCL	<35	
				Intermediate plasticity fines	SCI	35–50	
				High plasticity fines	SCH	50–70	
				Very high plasticity fines	SCV	70–90	
				Extremely high plasticity fines	SCE	>90	

Gravels (>50% of coarse material is of gravel size – >2 mm)

Sands (>50% of coarse material is of sand size – 0.06 mm to 2.00 mm)

Coarse soils (<35% fines)

Source: Courtesy of BS 5930–1981.

Table 9.9   Classification in EN ISO 14688-2.

Parameter	Terms	Definitions and Comment	Relevant Section
Nature of Soil			
Particle size fractions	Terms for low, medium, and high boulder and cobble content	Classification of very coarse soils only is provided. Classification of finer soils is based on plasticity and grading.	4.3, 9.1
Particle size grading based on shape of grading curve	Multi-graded	$c_u > 15, 1 < c_c < 3$	
	Medium graded	$6 < c_u < 15, c_c < 1$	
	Even graded	$c_u < 6, c_c < 1$	
	Gap graded	c_u high, c_c any value (c_u = uniformity coefficient, c_c = coefficient of curvature)	
Plasticity	Terms of low, intermediate, and high plasticity without definition	Terms are based on plastic limits so there is conflict with those based on plasticity chart and thus liquid limit (see below)	4.7
Organic content	Low organic	Organic content 2–6%	13.1
	Medium organic	Organic content 6–20%	
	High organic	Organic content > 20%	
State of soil			
Density index	Very loose	Density index < 15%	
	Loose	Density index 15–35%	
	Medium dense	Density index 35–65%	
	Dense	Density index 65–85%	
	Very dense	Density index > 85%	
		No term provided for classification based on SPT N values.	
Strength of fine soils	Extremely low	$c_u < 10$ kPa	5.3
	Very low	$c_u < 10$–20 kPa	
	low	$c_u < 20$–40 kPa	
	Medium	$c_u < 40$–75 kPa	
	High	$c_u < 75$–150 kPa	
	Very high	$c_u < 150$–200 kPa	
	Extremely high	$c_u > 300$ kPa	

(*Continued*)

Table 9.9   (*Continued*)

Parameter	Terms	Definitions and Comment	Relevant Section
Consistency Index	Very soft	Consistency index < 0.25	5.2
	Soft	Consistency index = 0.25–0.5	
	Firm	Consistency index = 0.50–0.75	
	Stiff	Consistency index = 0.75–1.00	
	Very Stiff	Consistency index > 1.00	
Other suitable parameters may be used	Dry density	No definition or terms are provided for these additional classification options	
	Clay activity		
	Mineralogy		
	Saturation index		
	Permeability		
	Compressibility		
	Swelling		
	Carbonate content		

Source: Norbury (2010).

Table 9.10   Brief summary of components with order of soil description.

Sr No	Description	
	Canada	UK
1	SOIL NAME	CONSISTENCY
2	Secondary Constituents	Color
3	Color Consistency	Secondary Constituents
4	Other content	Grain Size
5		Grain shape
6		Soil Name

of groundwater and/or piezometric levels. Two-dimensional cross sections can provide lateral variation of the soil and groundwater profile along the selected sections.

9.4 Fence Diagram

To visualize the sub-surface conditions in 3-dimensional view and see the ground variation in all directions, a fence diagram is usually built based on the borehole logs available. Figure 9.6 shows a fence diagram built using borehole data.

B Bo & Associates Inc.

METRIC												Page 1 of 1

RECORD OF BOREHOLE No W-01(MW1)

W.P.:	GS-20201001	LOCATION:	4834446.00N, 633301.00E	ORIGINATED BY: Steven
HWY:	A	BOREHOLE TYPE:	Rotary Mud Drilling	COMPILED BY: Carey
DATUM:	Geodetic	DATE:	2020.01.06 - 2020.01.09	CHECKED BY: MWB

ELEV DEPTH	DESCRIPTION (SOIL PROFILE)	STRAT PLOT	NUMBER	TYPE	"N" or RQD	GROUNDWATER CONDITIONS	ELEVATION SCALE	DYNAMIC CONE PENETRATION RESISTANCE PLOT / SHEAR STRENGTH (kPa)	WATER CONTENT (%)	UNIT WEIGHT γ kN/m³	GR	SA	SI	CL
142.0	Natural ground surface													
0.0 141.2	FILL, Silty and gravelly clay, brown.		AS1	AS										
0.8	SAND, some clay and silt, dark brown, compact. Presence of organic matter.		01	SS	10									
			02	SS	14		140				1	82		17
			03	SS	23		139							
			04	SS	23		138				4	84		12
			05	SS	14		137							
			06	SS	18		136 135							
134.5 7.5	CLAY, silty, traces of sand, grey, very soft to firm.		07	SS	0		134							
			08	SS	0		133 132							
			09	SS	4		131 130							
			10	SS	0		129							
			11A	SS	5		128				0	65		35
			11B	SS	7		127							
			12	SS	0		126							
			13	SS	0		125 124							
			14A	SS	0		123							
			14B	SS	8									
			15	SS	26		122							
			16	SS	91		121 120							
			17A	SS	73		119				12	69		19
			17B	SS	73		118							
			18	SS	76		117							
116.0 26.0	SAND, some clay and silt, traces of gravel.		19	SS	100		116 115				1	85		14
114.5 27.5	SAND, Clayey and silty, greyish brown, very dense.		20	SS	100		114				0	62		38
113.0 29.0	GRAVEL, some clay, grey, very dense.		21	SS	100		113 112							
111.5 30.5	END OF BOREHOLE		22	SS	100									

Water level 140.93 m on 2020-01-10

C:\Users\Win Myint Than\Desktop\Borehole Log\Style

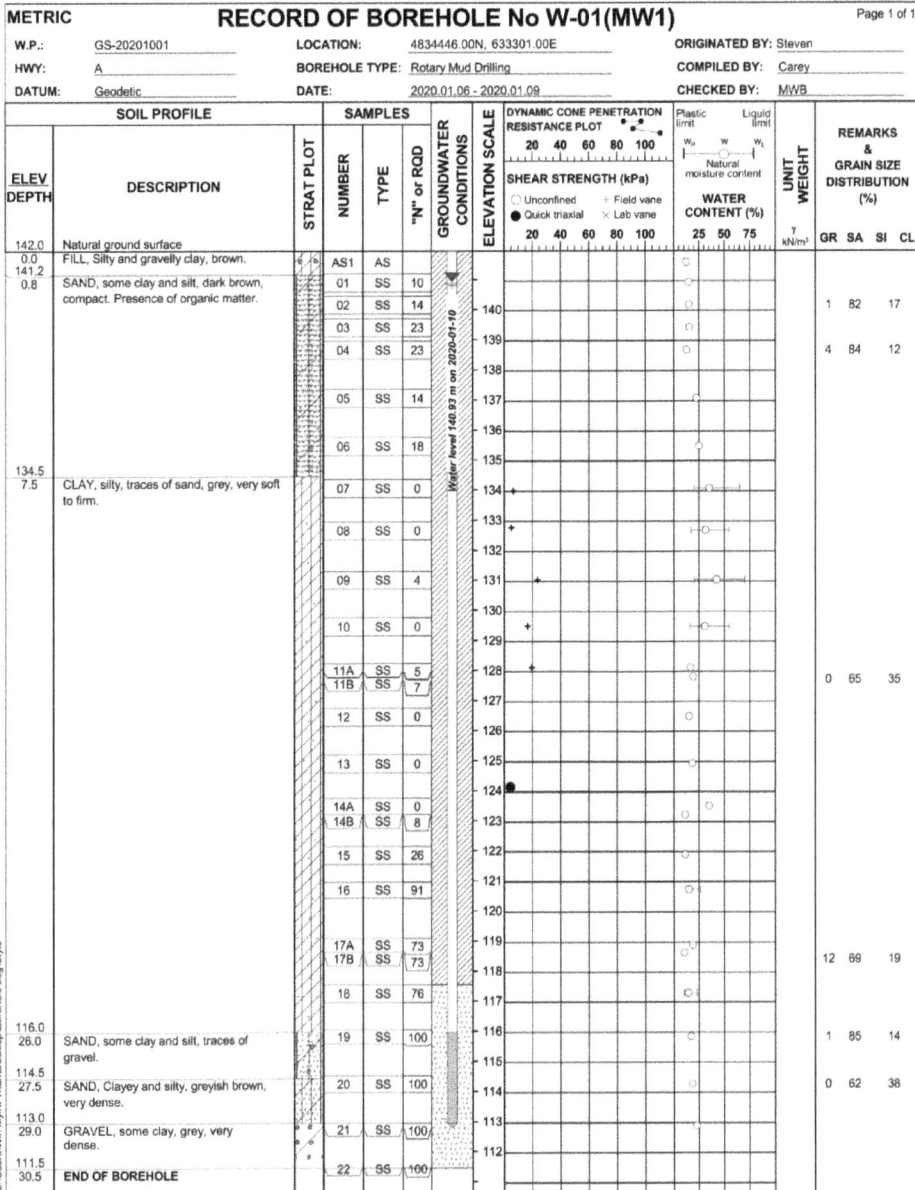

Figure 9.4   Typical borehole log presentation (Courtesy of Bo & Associates Inc.).

9.5 Ground Model

Ground models are usually built for slope stability analyses and stress deformation analyses. A ground model will consist of a two to three-dimensional ground

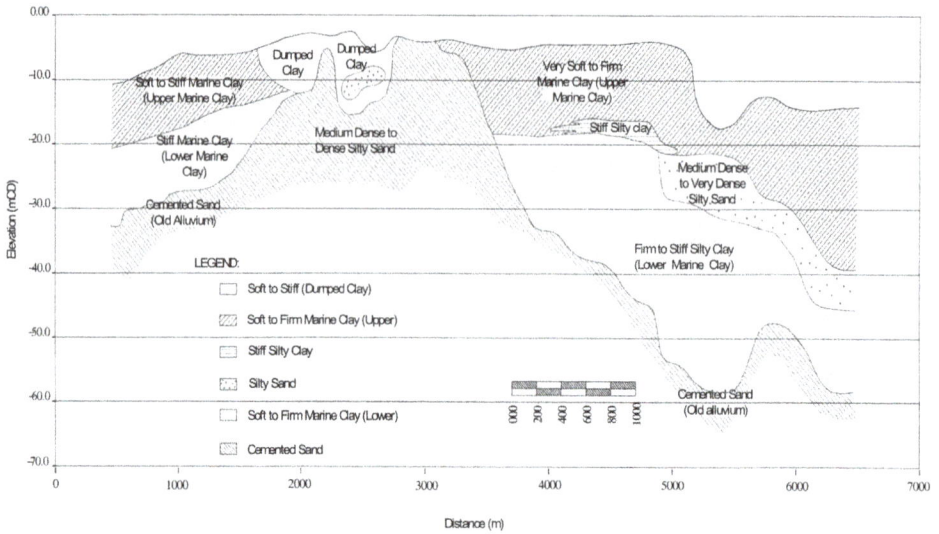

Figure 9.5   Soil profile cross section created along selected sections using borehole data.

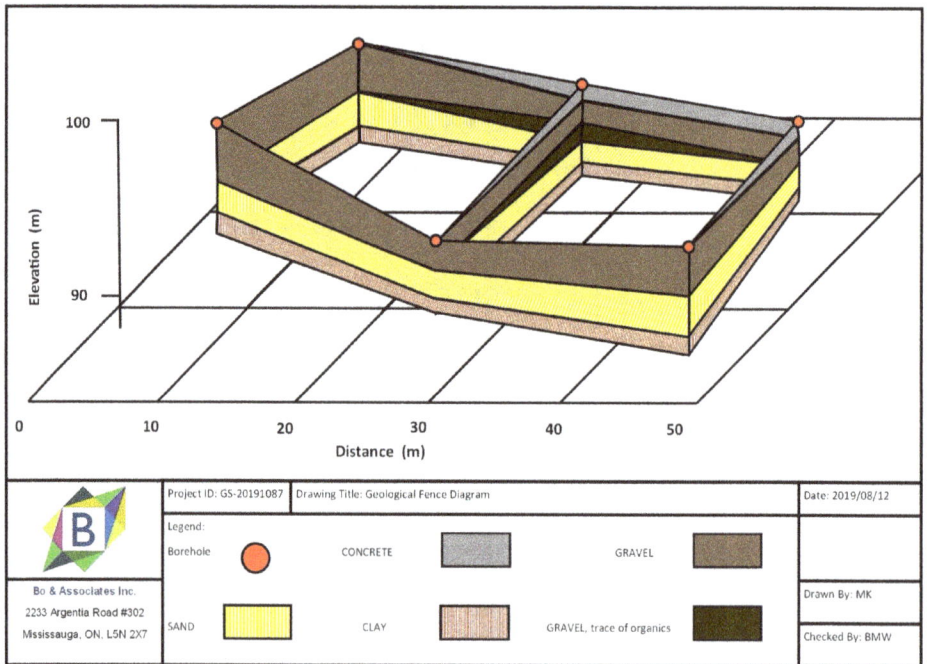

Figure 9.6   Generated fence diagram built using borehole data.

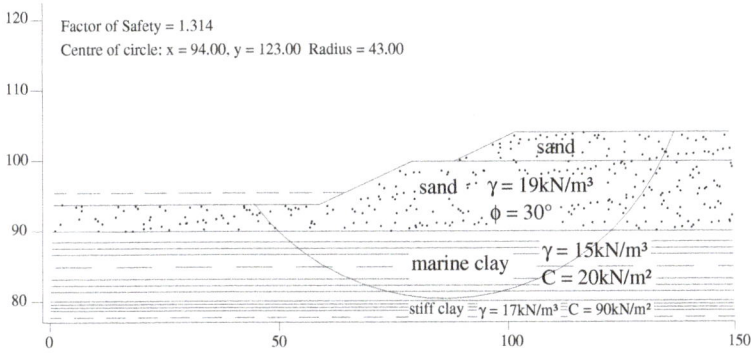

Figure 9.7   Ground model for slope stability analyses.

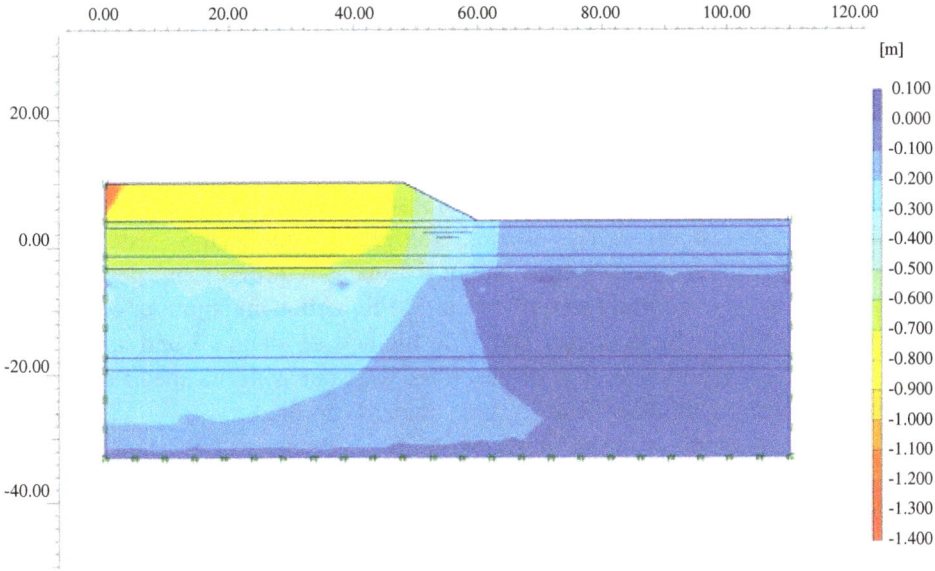

Figure 9.8   Stress deformation analyses applying Finite Element Modeling.

profile, groundwater and piezometric head profile, soil mechanical properties, and strength and deformation parameters of each formation and their variation with depth. Figures 9.7 and 9.8. show a ground model for slope stability analyses and stress deformation analyses, respectively, applying Finite Element Modeling.

9.6 Factual Reporting

Factual reporting of ground investigation data is conducted using data collected during intrusive ground investigations such as test pitting, borehole drilling, *in-situ* testing, laboratory testing, and installation. In this stage, collected information and test results are not required to be interpreted to drive geotechnical parameters required for design and analysis purposes. Presentation of factual findings is sufficient.

A factual report should consist of the following sections:

a. *Introduction*: Introductory statements describing the project name, type of development, location, name of the client's organization, and person who authorized the work should be stated in this section. The scope of work detailing items of work to be carried out and delivered should also be described in this section.

b. *Background*: In this section, the purpose of the ground investigation, any previous ground investigations, and any other related works carried out in the past should be described. In addition, general and brief descriptions of site conditions such as general topography, environment and surrounding infrastructures, and developments as well as access conditions should be described in this section. If there is any special condition of the site which could affect the ground investigation, it should also be mentioned in this section.

c. *Method of Investigation*: The method of investigation, such as test pitting or method of drilling and drilling technique used, is required to be described as information. The method and type of sample collection and number of samples collected should also be described. The method of standard and specialized *in-situ* tests performed should be described together with the specifications of the equipment, method of testing, and standards applied.

d. *Regional Geology*: A brief description of regional geology and its brief succession should be described based on the available existing literature.

e. *Site-Specific Ground Conditions:* A general profile of the sub-surface ground should be described based on the sub-surface conditions obtained from the boreholes investigated.

The characteristic of each individual soil layer should be described with depth and elevation of the top and bottom of the formation together with a range of thickness measured in each borehole. The types of soils encountered should be described with their color, condition of moisture, and consistency based on strength and compactness.

f. *Site-Specific Groundwater conditions*: Groundwater encountered during drilling, details of well installation for monitoring purposes, groundwater levels measured after 24 hours of installation, and subsequent monitoring data should be presented. A typical example of a well installation can be found in Chapter 8. Groundwater levels should be presented in depths below existing ground levels as well as their elevations.

The type of hydraulic conductivity tests carried out, the method of testing in compliance with specific standards, and the measurement details should also be presented.

g. *Laboratory tests performed and the results*: The types and number of tests performed and their results should be presented with any graphical presentation required, together with the information regarding standard complied with. Typical examples of presentation of grain size distributions and Atterberg limits test results are shown in Figures 9.9 and 9.10.

9.7 Interpreted Reporting

An interpreted report is designed to present interpreted data of measured *in-situ* and laboratory testing information in order to derive the characteristic values of geotechnical parameters and to provide geotechnical design recommendations for the project.

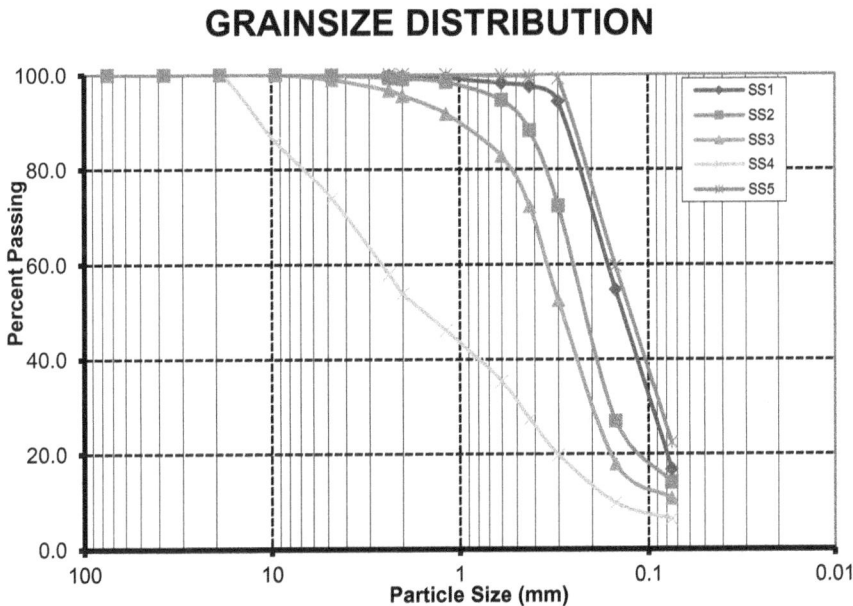

Figure 9.9   Presentation of grain size distributions.

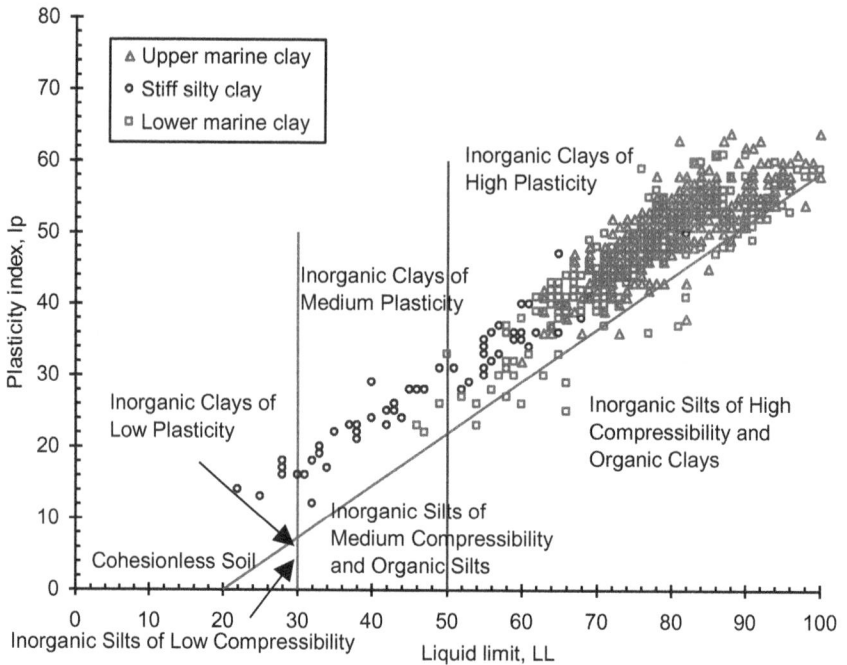

Figure 9.10   Presentation of Atterberg limits test results.

Some of the major parameters required for most geotechnical projects are as follows:

a. Strength Parameters
b. Stiffness, Compressibility, and Consolidation Parameters
c. Earth Pressure Coefficients
d. Hydraulic Conductivity Parameters

9.7.1 *Strength parameters*

Strength parameters are required for all soil layers encountered, which are to be used as input data for bearing capacity analyses and slope, excavation and retaining structure stability analyses. On the other hand, undrained parameters such as undrained shear strength (s_u) are required for cohesive soils and drained parameters such as internal friction angle (ϕ) are required for granular soils for bearing capacity analysis purposes.

Undrained strength parameters are normally obtained from direct measurements using FVT or laboratory measurements on the collected undisturbed sample applying the Unconsolidated Undrained (UU) test. More accurate Undrained

Shear Strength Parameters can be obtained by carrying out a Consolidated Undrained Test on the collected undisturbed sample. As undrained shear strength generally increases with the overburden stress, the rate of increase with depth or overburden stress should be presented as shown in Figure 9.11.

Drained Strength Parameters (ϕ') for granular soil could be interpreted from SPT test applying the equation provided in Chapter 6. Drained parameters for cohesive soils which consist of a drained friction angle only for normally consolidated soil could be obtained from the laboratory Consolidated Drained Test or Consolidated Undrained Test with pore pressure measurements of the collected undisturbed sample. Alternatively, the approximate drained internal friction angle for normally consolidated soil could be interpreted applying the relationship given by Kenny (1959) in Figure 9.12: For the strength of overconsolidated cohesive soils, which consists of both apparent cohesion (c') and drained friction angle (ϕ'), a set of parameters can only be obtained by carrying out laboratory tests on either a consolidated drain test or a consolidated undrained test with pore pressure measurement of the collected undisturbed samples.

Table 9.11 shows a typical presentation of strength parameters for soil formation.

9.7.2 *Stiffness, compressibility, and consolidation parameters*

Stiffness and compressibility parameters are required for settlement estimation of the foundation. Stiffness parameters such as Elastic Modulus are required for immediate settlement of the foundation on granular soils, whereas compressibility

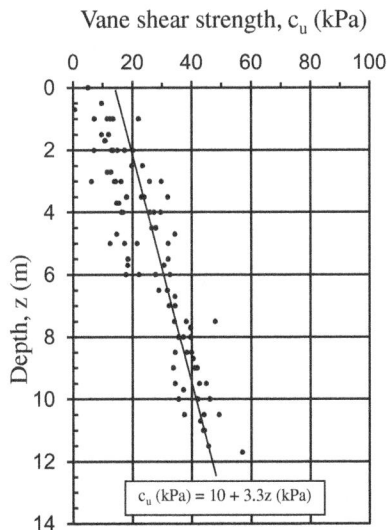

Figure 9.11   Shear Strength Vs Depth.

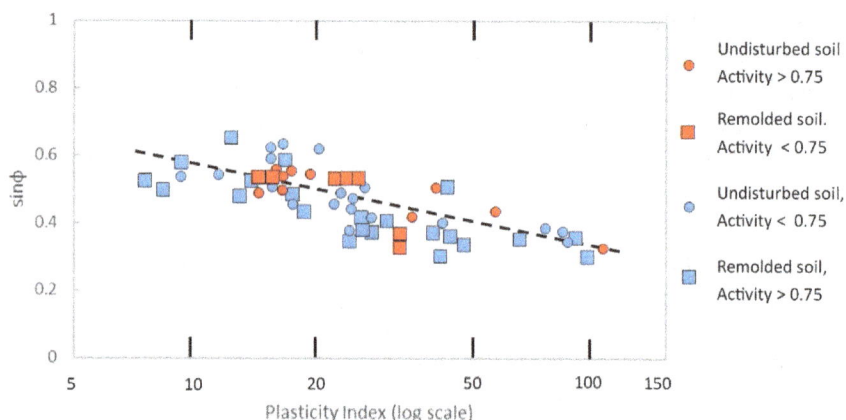

Figure 9.12   Approximate drained internal friction angle for normally consolidated soil.
Source: Kenney (1959).

Table 9.11   Typical presentation of strength parameters for soil formation.

Material	Unit Weight (kN/m³)	Void Ratio (e)	Undrained Shear Strength (Cu) (kPa)	Drained Cohesion (c') (kPa)	Drained Friction Angle (Φ') (Degrees)
Proposed New Fill	20	0.64	–		(32)
Existing Sand and Silt Fill	19	0.64	–		29–37 (30)
Medium Plasticity Clay	20	0.56	20	5	25
Native Silt and Sand	19	0.24	–		32–40 (30)

parameters are required for long-term settlement of the foundation caused by underlying cohesive and organic soils.

The elastic modulus (E) of granular soil could be correlated from the SPT blow counts obtained during ground investigation. Figure 9.13 show a correlation between Elastic Modulus and SPT values (Behpoor and *Ghahraman* 1989). More accurate Modulus parameters can be obtained from the laboratory tests carried out on the reconstituted sample in the laboratory.

Compressibility and consolidation parameters such as Compression index (C_c), re-compression index (C_r), secondary compression index (C_a), Coefficient of Consolidation (C_v), and Coefficient of Consolidation in recompression range (C_{vr})

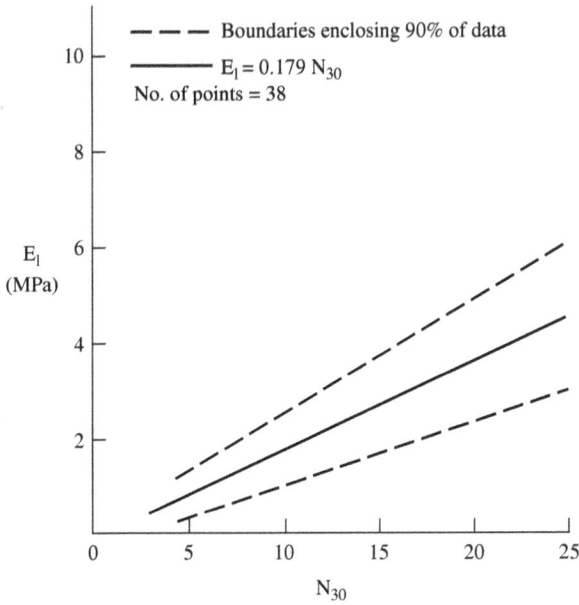

Figure 9.13 Correlation between elastic modulus and SPT values.

Source: Behpoor and Ghahraman (1989).

can be obtained from laboratory standard consolidation tests (BS 1377 Part 6, ASTM D 2435M-03).

Table 9.12 shows a typical presentation of compressibility and consolidation parameters for cohesive soils.

9.7.3 *Earth pressure coefficients*

Earth pressure coefficients are required when lateral earth pressures are required to be estimated for design of the retaining structures. Most built-up retaining structures are usually backfilled with suitably selected granular soils; therefore, most geotechnical reports provide earth pressure coefficients for granular backfill soils. It is a common practice to provide the Coefficient of Earth Pressure at rest (K_0), assuming the retaining structure is rigid and not allowing any soil mobilization as it is the conservative approach. An earth pressure coefficient at rest could be correlated from the drained friction angle of granular soil applying the following equation:

$$K_0 = 1 - \sin \phi \tag{9.1}$$

Table 9.12   Presentation of compressibility and consolidation parameters for cohesive soils.

Material	Initial Void Ratio	Compression Index	Recompression Index	Pre-consolidation Pressure (kN/m^2)	Coefficient of Consolidation (m^2/yr)	Hydraulic Conductivity (m/s)
Medium Plasticity Clay	0.56	0.24	0.03	75.5	1.2	1.18×10^{-10}

Where the flexible retaining structure is required to be designed allowing mobilization for both Active Earth Pressure (K_a) and Passive Earth Pressure (K_p) for granular soils. These parameters are correlated from the internal friction angle of soil applying the following equations:

$$K_a = 1 - \sin \phi \, / \, 1 + \sin \phi \tag{9.2}$$

$$K_p = 1 + \sin \phi \, / \, 1 - \sin \phi \tag{9.3}$$

Where the retaining structure is to be built by penetration of the flexible structure, such as sheet pile wall, into the existing natural soil formations, earth pressure coefficients of each formation should be provided based on the type of soil expected to be retained. Details of earth pressure coefficient calculations for various conditions of pressure, types of soils, and their over consolidation ratios can be found in many advanced soil mechanics textbooks (Das 1983). Typical presentations of earth pressure coefficient parameters are shown in Table 9.13.

9.7.4 *Hydraulic conductivity*

The hydraulic conductivity of water-bearing formations should be reported where excavation is required below the groundwater level and significant dewatering is necessary. Results from hydraulic conductivity tests carried out in accordance with the methods described in Section 4.6.3 should be reported in the interpreted report. A typical presentation of hydraulic conductivity test results together with soil formation details and the test section details should be provided as shown in Table 9.14.

Table 9.13   Typical presentation of earth pressure coefficient parameters.

Soil Type	Depth, m	Unit Weight, kN/m^3	C, kPa	$\varphi°$	K_0	K_a	K_p
Sand	1–3	19	–	(30)* 28–42	0.50	0.30	3.33

Note: *recommended design parameter.

Table 9.14   Typical presentation of hydraulic conductivity test results.

	BH2/MW2	BH4/MW4	BH5/MW5
Formation Tested	Clay and Silt	Clay and Silt	Clay and Silt
Tested Sections (m)	1.5–4.5	1.5–4.5	1.5–4.5
Hydraulic Conductivity (m/s)	3.6×10^{-7}	1.7×10^{-7}	8.3×10^{-7}

9.8 Design Recommendation

Details of the design process and reporting are beyond the scope of this textbook.

References

Aas, G., Lacasse, S., Lunne, T. and Heg, K. (1986). Use of in-situ tests for foundation design on clay. *Proceedings of 14th ASCE Specialty Conference on Use of In-situ Tests in Geotechnical Engineering,* Blackburg, U.S.A., pp. 1–30.

Amoroso, S., Monaco, P., Rollins, K. M., Holtrigter, M. and Thorp, A. (2015). Liquefaction assessment by seismic dilatometer test (SDMT) after 2010–2011 Canterbury earthquakes (New Zealand). *Proceedings, 6ICEGE Conference,* Christchurch, New Zealand.

Amoroso, S., Comina, C., Foti, S. and Marchetti, D. (2016). Preliminary results of P-wave and S-wave measurements by seismic dilatometer test (SPDMT) in Mirandola (Italy). In Lehane, B., Acosta-Martinez, H. E. and Kelly, R. (Eds.), *Proceedings, ISC'5 Conference.* Sydney: Australian Geomechanics Society, pp. 825–830.

Arabe, L. C. G. (1995). Application of in-situ test for evaluating geotechnical properties of quaternary clay deposit and residual soils. Ph.D. Thesis, Civil Engineering Department, PUC-RJ, June 1, 546 (in Portuguese).

Arulrajah, A., Nikraz, H. and Bo, M. W. (2004a). Observational methods of assessing improvement of marine clay, ground improvement. *Proceedings of the Institution of Civil Engineers (UK),* 8(4): 151–169.

Arulrajah, A., Bo, M. W., Nikraz, H. and Hashim, R. (2004b). Piezocone dissipation testing of Singapore marine clay at Changi, Geotechnical engineering. *Journal of the Southeast Asian Geotechnical Society,* 35(3): 119–126.

Arulrajah, A., Nikraz, H. and Bo, M. W. (2006a). Assessment of marine clay improvement under reclamation fills by in-situ testing methods. *Geotechnical and Geological Engineering, an International Journal, Kluwer Academic Publishers,* 24(1): 219–226.

Arulrajah, A., Bo, M. W., Nikraz, H. and Hashim, R. (2006b). Pre-reclamation in-situ testing of soft soil. *Australian Geomechanics: Journal of the Australian Geomechanics Society,* 41(4): 57–68.

Arulrajah, A., Bo, M. W. and Nikraz, H. (2008). Factors affecting prediction by the Asaoka and piezometer assessment methods, Geotechnical and geological engineering. *An International Journal*. Springer Publishers., 26(4): 417–430.

Arulrajah, A., Bo, M. W. and Nikraz, H. (2009). In-situ dissipation testing of soft soil under reclamation fills. *Australian Geomechanics: Journal of the Australian Geomechanics Society*, 44(1): 69–79.

Arulrajah, A., Bo, M. W., Piratheepan, J. and Disfani, M. M. (2011). In-situ testing of soft soil at a case study site with the self-boring pressuremeter. *ASTM Geotechnical Testing Journal*, 34(4): 355–363.

ASCE. (1972). Subsurface investigation for design and construction of foundations of buildings. Task Committee for Foundation Design Manual.

ASTM D 6089-97. (2003). Guide for Documenting a Ground-water Sampling Event. Standard D 6089. West Conshohocken, PA: ASTM International.

ASTM D 6564-00 (2005). Guide for Field Filtration of Ground-Water Samples. Standard D6564. West Conshohocken, PA: ASTM International.

ASTM D1586. (2018). Standard Test Method for Standard Penetration Test (SPT) and split-barrel sampling of soils. Standard D1586. West Conshohocken, PA: ASTM International.

ASTM D2113-02. (2006). Standard practice for rock core drilling and sampling of rock for site investigation. Standard D2113-06. West Conshohocken, PA: ASTM International.

ASTM D2216-19. (2019). Standard test method for laboratory determination of water (Moisture) content of soil and rock by mass. Standard D2216-19. West Conshohocken, PA: ASTM International.

ASTM D2435-03. (2003). Standard test method for one-dimensional consolidation properties of soils. Standard D2435-03. West Conshohocken, PA: ASTM International.

ASTM D2487-17e1. (2017). Standard practice for classification of soils for engineering purposes (Unified Soil Classification System). Standard D2487-17e1. West Conshohocken, PA: ASTM International.

ASTM D2573-08. (2008). Standard test method for field vane shear test in cohesive soil. Standard D2573-08. West Conshohocken, PA: ASTM International.

ASTM D2922. (2001). Standard test methods for density of soil and soil-aggregate in place by nuclear methods (shallow depth). Standard D2922. West Conshohocken, PA: ASTM International.

ASTM D2922-05. (2005). Edition, November 1, 2005 — Standard test methods for density of soil and soil-aggregate in place by nuclear methods (Shallow Depth). Standard D2922-05. West Conshohocken, PA: ASTM International.

ASTM D2974-20e1. (2020). Standard test methods for determining the water (moisture) content, ash content, and organic material of peat and other organic soils. Standard D2974-20e1. West Conshohocken, PA: ASTM International.

ASTM D3017. (2005). Standard test method for water content of soil and rock in place by nuclear methods (shallow depth). Standard D3017. West Conshohocken, PA: ASTM International.

ASTM D3213-13. (2013). Standard practices for handling, storing, and preparing soft intact marine soil. Standard D3213. West Conshohocken, PA: ASTM International.

ASTM D4220-14. (2014). Standard practices for preserving and transporting soil samples. Standard D4220-14. West Conshohocken, PA: ASTM International.

ASTM D4318-17e1. (2017). Standard test methods for liquid limit, plastic limit, and plasticity index of soils. Standard D4318-17e1. West Conshohocken, PA: ASTM International.

ASTM D4719-20. (2020). Standard test methods for prebored pressuremeter testing in soils. Standard D4719-20. West Conshohocken, PA: ASTM International.

ASTM D5778. (2012). Standard test method for electronic friction cone and piezocone penetration testing of soils. Standard D5778. West Conshohocken, PA: ASTM International.

ASTM D6151M. (2015). Standard practice for using hollow-stem augers for geotechnical exploration and soil sampling. Standard D6151M. West Conshohocken, PA: ASTM International.

ASTM D6391-11. (2011). Standard test method for field measurement of hydraulic conductivity using borehole infiltration. Standard D6391-11. West Conshohocken, PA: ASTM International.

ASTM D6517-00. (2000). Standard guide for field preservation of ground-water samples. Standard D6517-00. West Conshohocken, PA: ASTM International.

ASTM D6635-15. (2015). Standard test method for performing the flat plate dilatometer. Standard D6625-15. West Conshohocken, PA: ASTM International.

ASTM D6726. (2007). Standard guide for conducting borehole geophysical logging — electromagnetic induction. Standard D6726. West Conshohocken, PA: ASTM International.

ASTM D6727. (2007). Standard guide for conducting borehole geophysical logging — neutron. Standard D6727. West Conshohocken, PA: ASTM International.

ASTM D6951-03. (2003). Standard test method for use of the dynamic cone penetrometer in shallow pavement applications. Standard D6951-03. West Conshohocken, PA: ASTM International.

ASTM D7918-17a. (2017). Standard test method for measurement of flow properties and evaluation of wear, contaminants, and oxidative properties of lubricating grease by die extrusion method and preparation. Standard D7918-17a. West Conshohocken, PA: ASTM International.

ASTM D6913 / D6913M-17. (2017). Standard test methods for particle-size distribution (gradation) of soils using sieve analysis. Standard D6913 / D6913M-17. West Conshohocken, PA: ASTM International.

ASTM D7069-04. (2015). Standard guide for field quality assurance in a groundwater sampling event. West Conshohocken, PA: ASTM International.

ASTM D7400. (2019). Standard test methods for downhole seismic testing. Standard D7400. West Conshohocken, PA: ASTM International.

Baligh, M. M. and Levadoux, J. N. (1980). Pore pressure dissipation after cone penetration. Research Report. Cambridge, Massachusetts: Department of Civil Engineering, Massachusetts Institute of Technology. Publication No. R80-11, Order No. 662.

Baligh, M. M. (1985). The strain path method. *ASCE*, 111(GT9): 1108–1136.

Baligh, M. M., Azzouz A. S. and Chin, C. T. (1987). Disturbance due to ideal tube sampling. *ASCE*, 113(GT7): 739–757.

Battaglio, M., Bruzzi, D., Jamiolkowski, M. and Lancellotta, R. (1986). Interpretation of CPT's and CPTU's, part 1: Undrained penetration of clays. *Proceedings 4th International Geotech. Seminar on Field Instrumentation and In-situ Measurements*. Singapore: Nanyang Technological Institute.

Behpoor, L. and Ghahramani, A. (1989). Correlation of SPT to strength and modulus of elasticity of cohesive soils. *12th International Conference on Soil Mechanics and Foundation Engineering (Rio de Janerio)*. Rio de Janerio, Brazil.

Bjerrum, L. (1972). Embankments on soft ground. *Proceedings of the ASCE Conference on Performance of Earth-Supported Structures, Purdue University*, 2: 1–54.

Bo, M.W., Arulrajah, A. and Choa, V. (1997). Assessment of degree of consolidation in soil improvement project; International Conference on Ground Improvement Techniques, Macau. CI Premier Pte. Ltd., Singapore. pp. 71–80.

Bo, M. W., Arulrajah, A. and Choa, V. (1998a). The hydraulic conductivity of Singapore marine clay at Changi. *Quarterly Journal of Engineering Geology*, 31: 291–299.

Bo, M. W., Arulrajah, A., Choa, V. and Chang, M. F. (1998b). Site characterization for a land reclamation project at Changi in Singapore. *1st International Conference on Site Characterization (Geotechnical Site Characterization), April 1998, Robertson and Mayne Balkema*. Rotterdam, Atlanta, USA, pp. 333–340.

Bo, M. W, Chang, M. F., Arulrajah, A. and Choa, V. (2000). Undrained shear strength of the Singapore marine clay at Changi from in-situ tests, geotechnical engineering. *Journal of the Southeast Asian Geotechnical Society*, 31(2): 91–107.

Bo, M. W. and Choa, V. (2000). Site investigation practice in land reclamation project. *Year 2000 — Geotechnics Geotechnical Engineering Conference*, November 2000, Bangkok, Thailand.

Bo, M. W. and Choa, V. (2001). Various applications of cone penetration test in reclamation and soil improvement projects. *International Conference on In-Situ Measurement of Soil Properties and Case Histories*, May 2001, Bali, Indonesia, pp. 441–447.

Bo, M. W., Chu, J., Low, B. K. and Choa, V. (2003a). *Soil improvement — prefabricated vertical drain techniques*. Thomson Learning. Singapore.

Bo, M. W., Choa, V. and Hong, K. H. (2003b). Material characterization of Singapore marine clay at Changi. *Quarterly Journal of Engineering Geology and Hydrogeology*, 36(4): 305–319.

Bo, M. W. and Choa, V. (2004). *Reclamation and ground improvement*. Thomson Learning. Singapore.

Bo, M. W., Chu, J. and Choa, V. (2005). *Chapter 9 — The Changi East reclamation project in Singapore. Ground Improvement — Case Histories* by Indraratna, B. and Chu, J. (Eds.). Elsevier Ltd., Oxford, UK, pp. 247–276.

Bo, M. W., Chang, M.-F., Arulrajah, A. and Choa, V. (2012). Ground investigations for Changi East reclamation projects. *Geotechnical and Geological Engineering*, 30(1): 45–62.

Bo, M. W., Arulrajah, A., Leong, M., Horpibulsuk, S. and Disfani, M. M. (2014). Evaluating the in-situ hydraulic conductivity of soft soil under land reclamation fills with the BAT permeameter. *Engineering Geology*, 168: 98–103.

Bo, M. W., Arulrajah, A., Sukmak, P. and Horpibulsuk, S. (2015). Mineralogy and geotechnical properties of Singapore marine clay at Changi. *Soils and Foundations*, 55(3): 600–613.

Bo, M. W., Arulrajah, A., Choa, V., Horpibulsuk, S. and Samingthong, W. (March, 2017). Research-oriented ground investigation projects at Changi, Singapore. *Geotechnical Research*, 4(1): 30–46. DOI: 10.1680/jgere.16.00018.

Bo, M. W., Lwin, T. and Choa, V. (March, 2019). Application of specialised in situ tests in Changi East reclamation projects, Singapore. *Geotechnical Research*, 6(1): 3–18. DOI: 10.1680/jgere.18.00033.

Boussinesq, J. (1885). *Application des Potentiels a` l E'tude de l' E'qilibre et du Mouvement des Solides E'lastiques*. Paris: Gauthier-Villard.

British Standards Institution. (1990). BS 1377-Part 9: In-situ tests. *Methods of Test for Soils for Civil Engineering Purposes.*

British Standards Institution. (1990). BS 1377-1: General requirements and sample preparation. *Methods of Test for Soils for Civil Engineering Purposes.*

British Standards Institution. (1990). BS 1377-2: Classification tests. *Methods of Test for Soils for Civil Engineering Purposes.*

British Standards Institution. (1990). BS 1377-3: Chemical and electro-chemical test. *Methods of Test for Soils for Civil Engineering Purposes.*

British Standards Institution. (1981). BS 5930: 1981: *Code of practice for site investigation.*

British Standards Institution. (1999). BS 5930: 1999: *Code of practice for site investigation.*

Canadian Foundation Manual (2006). Canadian Geotechnical Society. Vancouver, BC, Canada. 4th Ed., p. 488.

Cao, L. F. (1997). Interpretation of in-situ test in clay with particular reference to reclaimed sites. Ph.D Thesis. Singapore: Nanyang Technological University.

Chandler, R. J., (1988). The in-situ measurement of the undrained shear strength of clays using the field vane. In Richards, A. F. (Ed.) *Vane Shear Strength Testing in Soils: Field and Laboratory Studies*, ASTM STP 1014. Philadelphia: American Society for Testing and Materials, pp. 13–44.

Chang, M. F., Choa, V., Cao, L. F. and Bo, M. W. (1997). Overconsolidation ratio of a seabed clay from in-situ test. *14th International Conference on Soils Mechanics and Foundation Engineering*, September 1997, Hamburg, Germany, pp. 453–456.

Chen, B. S-Y. and Mayne, P. W. (1996). Statistical relationships between piezocone measurements and stress history of clays. *Canadian Geotechnical Journal*, 33(3): 488–498.

Chu, J. (2002). Part 7 sampling methods & effect of sample disturbance [PowerPoint presentation]. August 2021.

Construction Health and Safety Manual. (2001). M029, Infrastructure Health & Safety Association, Ontario, Canada.

Coutinho, R. R., Oliveira, J. T. R. and Danziger, F. A. B. (December, 1993). Geotechnical characterization of a soft clay deposit at Recife, soils and rocks. *Journal of Brazilian Geotechnical Society*, 16(4): 255–266 (in Portuguese).

Das, B. M. (1983). *Advanced Soil Mechanics*. New York: McGraw-Hill.

Das, B. and Sivakugan, N. (2017). *Geotechnical Engineering: A Practical Problem Solving Approach*. Florida, USA: J. Ross Publishing Inc.

Das, B. and Sivakugan, N. (2017). *Fundamentals of Geotechnical Engineering*. 5th Ed. Boston, MA: Cengage Learning.

Décourt, L., de Camargo Barros, J. M., Gandolfo, O. C. B., Filho, A. R. Q. and Penna, F. D. (2016). Maximum shear modulus of a Brazilian lateritic soil from in situ and laboratory tests. In Lehane, B., Acosta-Martinez, H. E. and Kelly, R. (Eds.), *Proceedings, ISC'5 Conference*. Sydney: Australian Geomechanics Society, pp. 1417–1422.

Denver, H. (1988). CPT and shear strength of clay. In De Ruiter (Ed.), *Penetration Testing 1988, ISOPT-1*. Balkema: Rotterdam.

De Ruiter, J. (1982). The static cone penetration test: State-of-the-art report. *Proceedings of the Second European Symposium on Penetration Testing*, Amsterdam, The Netherlands, pp. 389–405.

Dobie, M. J. D. (1988). A study of cone penetration tests in Singapore marine clay. *Proceedings of the 1st International Symposium on Penetration Testing*. Orlando, USA, Vol. 2, pp. 737–744.

Eaton, G. P. and Watkins, J. S. (1967). The use of seismic-refraction and gravity methods in hydrogeological investigations, In Morey, L.W. (Ed)., *Mining and Ground Water Geophysics : Geological Survey of Canada Economic Geology Report 26*, pp. 544–568. Department of Energy, Mines, and Resources, Ottawa, Canada.

Eaton, G. P. and Watkins, J. S. (1970). The use of seismic-refraction and gravity methods in hydrologic investigations. In Morey, L. W. (Ed.), *Mining and Ground-water Geophysics, 1967: Geological Survey of Canada Economic Geology Report 26*, p. 772. Department of Energy, Mines, and Resources, Ottawa, Canada.

Fusao, O., Atsushi, Y., Tadashi, H. and Matsuo, A. (September, 1996). Application of laval type large diameter sampler to soft clay in Japan. *Soil and Foundations*, 36(3): 99–111. Japanese Geotechnical Society.

George, E. A. and Ajayi, L. A. (1995). National report on cone penetration testing, Nigeria. *International Symposium on Cone Penetration Testing*. Swedish Geotechnical Society, Vol. 1. National Report.

Gibson, R. E. (1963). An analysis of system flexibility and its effect on time-lag in pore-water pressure measurements. *Geotechnique*, 13: 1–11.

Google Earth Pro 7.3.4.8248. (July 16, 2021). Banff, Alberta. 51°10'41.29"N, 115°34'15.00"W, Eye alt 3 km. Maxar Technology. http://www.earth.google.com.

Google Earth Pro 7.3.4.8248. (July 16, 2021). Thunder Bay, Ontario. 48°23'24.10"N, 89°14'57.30"W, Eye alt 350 km. NOAA. http://www.earth.google.com.

Griffiths, D. H. and King, R. F. (1986). *Applied Geophysics for Engineers and Geologists*. 2nd Ed., Oxford, England: Pergamon Press.

Gwyn, Q. H. J., Fraser, J. Z. and Owen, N. (1975). Drift thickness of the Cornwall-Huntingdon Area, Southern Ontario; Ontario Div. Mines Prelim. Map P. 1013, Drift Thickness Ser., scale 1:50 000 Geology and compilation, 1974.

Haskell, N. A. (1953). The dispersion of surface waves on multilayered media. *Bulletin of Seismological Society of America*, 43: 17–34.

Hawkins, P. G., Mair, R. J., Mathieson, W. G. and Wood, D. M. (1990). Pressuremeter measurement of total horizontal stress in stiff clay. *Proceedings 3rd International Symposium on Pressuremeters*, Oxford University, UK, pp. 321–330.

Head, J. M. (1986). Planning and design of site investigations. In Hawkins, A. B. (Ed.), *Site Investigation Practice: Assessing BS 5930*. Geological Society, Engineering Geology Special Publication No. 2. London. UK.

Herrmann, R. B. (1973). Some aspects of band-pass filtering of surface waves. *Bulletin of Seismological Society of America*, 62(2): 663–671.

Hight, D. W. (1986). Laboratory testing: Assessing BS5930. *Proceeding 20th Regional Meeting of Engineering Group of Geological Society*. University of Surrey. (Presented in 1984). London, UK. Engineering Geology Special Publication No. 2. pp. 43–52.

Hight, D. W., Böese, R., Butcher, A. P., Clayton, C. R. I., and Smith, P. R. (1992). Disturbance of the Bothkennar Clay prior to laboratory testing. *Géotechnique*, 42(2), 199–217.

Hight, D. W., and Leroueil, S. (2003). Characterisation of soils for engineering purposes. *Characterisation and Engineering Properties of Natural Soils*, Tan et al. (eds.), Balkema, 1, 255–360.

Intergovernmental Committee on Surveying & Mapping. (No date). A stereo pair of aerial photographs[Artwork]. https://www.icsm.gov.au/education/fundamentals-mapping/surveying-mapping/introduction.

International Organization for Standardization. (2004). Geotechnical investigation and testing — Identification and classification of soil — Part 2: Principles for a classification. ISO/DIS Standard No. 14688-2.

Infrastructure Health & Safety Association. (2019). Construction health and safety manual. Canada.

Jones, S. R. (1995). Engineering properties of alluvial soils in newcastle using cone penetration testing. Engineering Geology of the Newcastle-Gosford Region, Aust Geomechanics Society.

Jorgensen, M. and Denver, H. (1992). CPT-interpretation. *11th Nordiske Geoteknikermode, DGF Bulletin 9*, Aalborg, Denmark (in Danish).

Kammer Mortensen, J., Hansen, G. and Sorensen, B. (1991). Correlation of CPT and field vane tests for clay tills. Danish Geotechnical Society, DGF Bulletin 7.

Karrow, P. F. and White, O. L. (1998). *Urban Geology of Canadian Cities*. Geological Association of Canada. Geological Association of Canada. Newfoundland, Canada.

Kenney, T. C. (1959). Discussion: Journal of the soil mechanics and foundation engineering division. *ASCE*, 85(SM3): 67–79.

Kessler, H., Mathers, S. and Sobisch, H. G. (2009). The capture and dissemination of integrated 3D geospatial knowledge at the British Geological Survey using GSI3D software and methodology. *Computers & Geosciences,* 35(6): 1311–1321.

Keys, W. S. and MacCary, L. M. (1985). Application of borehole geophysics to water-resources investigations. Fourth Printing, Techniques of Water-Resources Investigation of the United States Geological Survey, Department of the Interior, U.S Geological Survey.

Kjekstad, O., Lunne, T. and Clausen, C. J. F. (1978). Comparison between in-situ cone resistance and laboratory strength for overconsolidated north sea clays. Marine Geotechnology, No. 4, Also Norwegian Geotechnical Institute, Publication, p. 124.

Konrad, J. M. and Law, K. (1987). Preconsolidation pressure from piezocone tests in marine clays. *Geotechnique,* 37(2): 177–190.

Kristjansson, F. J. and Thorleifson, L. H. (1991). Surficial Geology, Beardmore-Geraldton, Ontario; Geological Survey of Canada, Map 1768A; Ontario Geological Survey, Map 2535, scale 1: 100 000.

Kulhawy, F. H. and Mayne, P. W. (1990). *Manual of Estimating Soil Properties for Foundation Design.* Ithaca, NY: Electric Power Research Institute.

La Rochelle, P., Sarraih, J., Tavenas, F., Roy, M. and Leroueil, S. (1981). Causes of sampling disturbance and design of a new sampler for sensitive soils. *Canadian Journal,* 18(1): 52–66.

La Rochelle, P., Zebdi, M., Leroueil, S., Tavenas, F. and Virely, D. (1988). Piezocone tests in sensitive clays of Eastern Canada. In De Ruiter (Ed.), *Penetration Testing 1988, ISOPT-1.* Rotterdam: Balkema.

Ladd, C. C. and DeGroot, D. J. (2003). Recommended practice for soft ground site characterization. Soil & Rock America 2003, (Proc. 12th Pan American Conf., MIT), Verlag Glückauf, Essen: 3–57.

Lefebvre, G. and Poulin, C. (1979). A new method of sampling in sensitive clay. *Canadian Geotechnical Journal,* 16(1): 226–233. DOI:10.1139/t79-019.

Lugeon, M. (1933). *Barrages et Geologie.* Paris: Dunod.

Lunne, T. and Kleven, A. (1981). Role of CPT in North Sea foundation engineering. *Proceeding ASCE National Convention at St. Louis Cone Penetration Testing and Experience.* Missouri, USA.

Lynch, E. J. (1962). *Formation Evaluation.* New York: Harper and Row.

Mair, R. J. and Wood, D. M. (1987). Pressuremeter testing: Methods and interpretation. CIRIA Ground Engineering Report: In-situ Testing, Butterworths, London.

Marchetti, S. (1980). In-situ tests by flat dilatometer. *Journal of Geotechnical Engineering Division, ASCE,* 106(GT3): 299–321.

Marchetti, S. and Crapps, D. K. (1981). *Flat Dilatometer Manual.* Gainesville, Florida, USA: Schmertmann and Crapps Inc., Consulting Geotechnical Engineers.

Marchetti, S. and Totani, G. (1989). Ch evaluation from DMTA dissipation curves. *Proceedings of the 12th International Conference on Soil Mechanics and Foundation Engineering, Rio de Janeiro,* 1: 281–286.

Marchetti, S., Monaco, P., Totani, G. and Marchetti, D. (2008). *In-situ Test by Seismic Dilatometer (SDMT)*. Geotechnical Special Publication. DOI:10.1061/40962(325)7.

Marchetti, D. (2019). Application and recent developments of the Flat Dilatometer (DMT) and Seismic Dilatometer (SDMT). *Proceedings of the XVII ECSMGE-2019, Geotechnical Engineering Foundation of the Future*, pp. 1–10. London, UK.

Marchetti, D., Monaco, P., Amoroso, S. and Minarelli, L. (2019). In situ tests by Medusa DMT. *Proceedings of the XVII ECSMGE-2019, Geotechnical Engineering Foundation of the Future*, pp. 1–1. London, UK.

Marsland, A. and Randolph, M. F. (1977). Comparison of the results from pressuremeter tests and large in-situ plate tests in London clay. *Geotechnique*, 27(2): 217–243.

McMechan, G. A. and Yedlin, M. J. (1981). Analysis of dispersive waves by wave field transformation. *Geophysics*, 46(6): 869–874.

Miller, J. A. (1990). Ground Water Atlas of the United States: Segment 6, Alabama, Florida, Georgia, South Carolina, Hydrologic Atlas, Series No. 730-G. Florida, USA: U.S. Geological Survey. DOI:10.3133/ha730G.

Ministry of Energy, Northern Development and Mines. (June 06, 2021). Abandoned mines database; OGSEarth online database. https://www.mndm.gov.on.ca/en/mines-and-minerals/applications/ogsearth/abandoned-mines.

Ministry of Transportation Ontario. (June, 2013). Provincial pavement engineering investigation guidelines version 1.1.

Ministry of the Environment, Conservation and Parks. (2021). Well records database. Ontario: Ministry of Ontario.

Na, Y. M., Choa, V., Chang, M. F., Teh, C. I. and Bo, M. W. (August, 1999). Estimation of geotechnical parameters of granular soils from various in-situ tests. In Hong *et al.* (Eds.), *11th Asian Regional Conference on Soil Mechanics and Geotechnical Engineering*. Rotterdam, Seoul, Korea: Balkema, pp. 277–280.

National Collection of Aerial Photography. (No date). Stereoscope [Photograph]. https://ncap.org.uk/sites/default/files/stereoscope.jpg.

Nhuan, B. D. and Tuong, D. T. (1985). Some result from study on soil investigation by Swedish equipment. IBST-SGI report.

Norbury, D. (2010). *Soils and Rocks Description in Engineering Practice*. Whittles Publishing; CRC Press; Taylor & Francis Group. Scotland, UK.

Oka, F., Yashima, A., Hashimoto, T. and Amemiya, M (1996). Application of Laval type large diameter sampler to soft clay in Japan. *Soils and Foundations*, 36(3): 99–111, September 1996, Japanese Geotechnical Society.

Ontario Geological Survey. (1977). Sudbury-Cobalt Geological Compilation Series, Algoma, Manitoulin, Nipissing, Parry Sound, Sudbury and Timiskaming District; Ontario Geological Survey, Final Map M.2361, scale 1:253 440.

Pan, X., Guo, W., Aung, Z., Nyo, A. K. K., Chiam, K., Wu, D. and Chu, J. (2018). Procedure for Establishing a 3D Geological Model for Singapore. In Shi X., Liu Z., Liu J. (Eds.) Proceedings of GeoShanghai 2018 International Conference:

Transportation Geotechnics and Pavement Engineering. GSIC 2018. Springer, Singapore. pp. 81-89. https://doi.org/10.1007/978-981-13-0011-0_9

Park, C. B., Miller, R. D. and Xia, J. (1997). Multichannel analysis of surface waves. Kansas Geological Survey. Open-file Report #97-10.

Pein, T., Monaco, P., Amoroso, S. and Marchetti, D. (2019). Comparisons of shear wave velocity measurements by SDMT and by other in situ techniques at well documented test sites. *Proceedings, 7ICEGE, Italian Geotechnical Society*, Rome.

Powell, J. J. M. and Quarterman, R. S. T. (1988). The interpretation of cone penetration tests in clays, with particular reference to rate effects. In De Ruiter (Ed.), *Penetration Testing 1988, ISOPT-1*. Rotterdam: Balkema.

Rad, N. S. and Lunne, T. (1988). Direct correlations between piezocone tests results and undrained shear strength of clay. *1st International Conference of Penetration Testing*, Orlando, Florida.

Robertson, P. K. and Campanella, R. G. (1983). Interpretation of cone penetration tests, part I: Sand. *Canadian Geotechnical Journal*, 20(4): 719–733.

Rocha Filho, P. and Alencar, J. A. (1985). Piezocone tests in Rio de Janeiro soft clay deposit. *Proceedings of the 11th International Conference on Soil Mechanics and Foundation Engineering*, San Francisco, Balkema, Rotterdam, Vol. 2, pp. 859–862.

Rocha-Filho, P. (August, 1987). Determination of the undrained shear strength of two soft clay deposits using piezocone tests. *International Symposium on Geotechnical Engineering of Soft Soils*, Mexico, pp. 125–120.

Sanglerat, G. (1972). *The Penetrometer and Soil Exploration*. Amsterdam: Elsevier, p. 464.

Schnaid, F., Belloli, M. V. A., Odebrecht, E. and Marchetti, D. (2018). Interpretation of the DMT in silts. *Geotechnical Testing Journal*, 41(5): 868–876.

Self, S. and Entwisle, D. C. (2006). The structure and operation of the BGS National Geotechnical Properties Database. British Geological Survey Internal Report, IR/06/092.

Self, S., Entwisle, D. and Northmore, K. (2012). The structure and operation of the BGS National Geotechnical Properties Database. Version 2.

Senneset, K., Janbu, N. and Svano, G. (1982). Strength and deformation parameters from cone penetration tests. *2nd European Symposium on Penetration Testing*, Amsterdam, pp. 863–870.

Singer, S. N., Cheng, C. K. and Scafe, M. G. (1997). *The Hydrogeology of Southern Ontario*. Toronto: Ontario Ministry of Environment and Energy.

Soares, M.M., Lunne, T., Almeida, M.S.S. and Danziger, F.A.B. (1986). Piezocone and dilatometer tests in a very soft Rio de Janeiro clay. Proc. Int. Symp. Geot. Eng. of Soft Soils, Mexico.

Soe Moe, K. W., Bo, M. W., Arulrajah, A., Horpibulsuk, S. and Hoy, M. (2020). Stiffness parameters from cone pressuremeter tests at Changi East, Singapore. *Proceeding of Civil Engineers, Ground Improvement*, UK.

Sowers, G. B. and Sowers, G. F. (1970). *Introductory Soil Mechanics and Foundations.* 3rd ed., New York: Macmillan Publishers.

Tanaka, H. (1994). Vane shear strength of Japanese marine clays and applicability of Bjerrum's correction factor. *Soils and Foundations*, 34(3): 39–48.

Terzaghi, K., Peck, R. B. and Mesri, G. (1996). *Soil Mechanics in Engineering Practice.* 3rd Ed., New York: John Wiley and Sons, Inc.

Thorburn, S. (1986). *Field Testing: The Standard Penetration Test.* London: Geological Society, Engineering Geology Special Publications, Vol. 2, pp. 21–26. DOI:10.1144/GSL.1986.002.01.07.

Torstensson, B. A. (1984). A new system for ground water monitoring. *Ground Water Monitoring Review*, 4(4): 131–138.

Torstensson, B. A. and Schellingerhout, A. J. G. (October, 1999). *Ground water monitoring with the BAT-system.* Geoteckniek. The Netherlands, pp. 1–13.

U.S. Navy. (1971). Soil mechanics, foundations and earth structures. NAVFAC Design Manual DM-7, Washington, D.C.

USBR. (1952). Progress report of research on the penetration resistance method of subsurface exploration. USBR Earth Lab, Rept. E-314, Rev 1955.

West, N. E. (1990). Structure and function of microphytic soil crust in wildland ecosystems of arid to semi-arid regions. *Advances in Ecological Research*, 20: 180–223.

Wilson, S., Card, G. and Haines, S. (2009). *Ground Gas Handbook.* UK: Whittles Publishing.

Wolff, T. F. (1995). *Spreadsheet Application in Geotechnical Engineering.* Boston, MA: PWS Publishing Company, A Division of International Thomson Publishing Inc.

Wroth, C. P. (December, 1984). The interpretation of in-situ test. 24th Rankine Lecture. *Geotechniques.* 34(4): 449–489; *Proceedings 20th Regional Meeting, Engineering Group, Geological Society*, Guildford 2, pp. 32–35.

Xia, J., Miller, R. D. and Park, C. B. (1999). Estimation of near-surface shear-wave velocity by inversion of Rayleigh waves. *Geophysics*, 64(3): 691–700.

Zohdy, A. A. R., Eaton, G. P. and Mabey, D. R. (1984). Application of surface geophysics to ground-water investigations. *Third Printing, Techniques of Water-Resources Investigation of the United States Geological Survey*, Department of the Interior, U.S. Geological Survey. Colorado, USA.

Index

A

active earth pressure, 228
aerial photography, 24–27
apparent cohesion, 225
aquifer, 14–18, 46, 81
archeological map, 2, 7
Artesian pressure, 195
Atterberg's limits, 154, 213–214, 223–224
Augering, 55, 57–58
Auto ram sounding, 115

B

balloon displacement method, 119–120
BAT permeameter, 116–119
bearing capacity, 36, 42, 224
bedrock drilling, 47
borehole, 38–45, 142–148
borehole data, 31
Boussinesq 1885 method, 46
Boyle's law, 117
British Soil Classification System (BSCS), 213
bulk density, 135–136, 142, 147–148

C

cable percussive drilling, 56–57
Casagrande open type piezometer, 192–195
compressibility, 4, 225–227
compression index, 226, 228

compressional and shear wave velocities, 139, 147
Cone Penetration Tests (CPTs), 89–90, 122–124
  overconsolidation ratio, 97
  undrained shear strength, 90–91, 94, 96
cone pressuremeter test (CPMT), 108–109
cone resistance, 89, 91, 96, 110–111, 120–121
consolidation parameters, 226–227
constant head test, 81
constrained modulus, 110
core cutter method, 119
coring, 62
crosshole seismic testing, 146, 147
cross sections, 215, 218

D

damping tube(s), 139
deep reference point, 184–185
degree of consolidation, 121–122, 124
degree of dissipation, 126
dewater(-ing), 38, 40, 228
desk study, 2, 7
dilatometer, 50, 101, 104, 122
dilatometer modulus, 100
dilatometer test (DMT), 124–125
dipole-dipole configuration, 130–131

dissipation curve, 98, 102–103, 106
dissipation test, 98, 103–104, 108, 122, 125–126
disturbed sample, 154
downhole seismic testing, 142–144, 146
down the hole hammer drilling, 62–64
drainage map, 7
drained friction angle, 50, 72–73, 225, 227
drained strength parameters, 225
drift thickness, 17, 21
drilling, 54–55
drilling methods
    augering, 57–58
    cable percussive, 56–57
    coring, 62
    down the hole hammer, 62–64
    hollow stem flight auger, 59
    rotary open hole drilling, 60
    wash boring, 56
drill rig mounting, 64–65
driller's log vs. engineer's logs, 213
dynamic cone testing, 113–114

E
earthquake, 38
earth pressure coefficients, 227–228
elastic modulus, 225–227
electric conductivity probe, 66
electrical survey, 17
electrode array, 130
electromagnetic survey, 17
embankment stability, 38
energy transfer ratio, 74
engineer's log, 213
engineering site investigation, 128
environmental investigation, 129
excessive pore pressure, 77

F
factual reporting, 5, 222
falling head test, 80
fence diagram, 218
field description, 209
field vane tests (FVTs), 75–77
finite element modeling, 221

flat dilatometer test (DMT), 99–101
    overconsolidation ratio, 101
    undrained shear strength, 101
flexible structure, 228
foundation
    culvert, 38, 42
    deep, 38, 40, 46
    shallow, 38, 45
foundation analysis, 142
frost penetration, 38, 47

G
gamma, 66, 135–136
Gamma–Gamma probe, 66–67, 135–138
gas migration, 206
gas monitoring well, 201–206
geochemistry, 21
geochronology, 17
geological map, 7–8, 14, 29–31, 45
3-D geological modeling
    building of, 33–34
    data preparation, 31–32
    fence diagram construction, 33
    interpreted cross section creation, 32–33
    procedures of, 30–31
    recent development, 29–30
    software packages, 30
geological modeling software, 30
geological sub-surface, 30–31
geology, 2, 8, 12–14, 17, 21–22, 29–30, 34, 222
geomorphological evolution, 30
geophone, 109, 112, 127–128, 143–144, 146, 149
geophysical investigation, 127
geophysical logging, 132–135
geophysical mapping, 37
geophysical survey, 4–5, 28–29
geotechnical instrumentation, 5
geotechnical parameter, 1, 4–5, 35, 37, 41–42, 48, 50, 72, 89–90, 99, 108–109, 111, 119, 153, 222–223
Gouda piezocone, 91
grade level, 46
ground investigation, 1

ground models, 219–221
groundwater pressure, 47, 53, 56–57,
  59–61
groundwater sampling, 158–160
groundwater survey, 129
grout mixture, 190
grouted (bore-)hole, 190

H
heaving, 38, 185
hollow stem flight auger, 59
horizontal stress index, 100
hydraulic balancing, 59
hydraulic conductivity, 81–88, 228–229
hydraulic gradient, 81
hydraulic pressure, 14, 176
hydraulic uplift, 46
hydrogeological maps, 7, 14, 16

I
inclinometer, 199–201
influence depth, 3, 45, 47, 120, 160
infrastructure development, 1–3, 23,
  36–37, 39, 45
in-situ testing, 3–5, 36–37, 49–50, 69, 89,
  99, 121–122, 127, 222
internal friction angle, 224–226, 228
interpreted reporting, 223–229
    earth pressure coefficients, 227–228
    hydraulic conductivity parameters,
      228
    stiffness, compressibility, and
      consolidation parameters, 225–227
    strength parameters, 224–225
intrusive ground investigation, 2–3, 35
    adopting strategy of, 36–38
    depth of investigation, 45–47
    in-situ testing, 49–50
    methods of drilling, 47–48
    monitoring and measurements, 51–52
    sampling methods, 48–49
    scope of, 35–36
    selection of locations of boreholes,
      38–45
    services and archaeological locations,
      51

L
laboratory test, 1, 3, 5, 36, 55, 103, 164,
  171, 180, 222–223
Lame's compressibility, 142, 148
landfill, 36, 39, 43–45, 47, 61, 120, 129
larger-diameter samples, 164–171
large-scale investigation, 36
lateral earth pressure, 107, 227
lateral stress index, 101
laval sampler, 165–167
lee partition configuration, 130
Light Detection and Ranging (LiDAR)
  survey, 2, 27–28
liquefaction assessment, 143
lithostratigraphic(-al), 33
lugeon value, 87–88

M
2:1 method, 46
made ground, 1, 45
magnetic survey, 17
material index, 100–102
membrane lift off pressure, xv, 106–107
membrane stiffness, 99
micro-tremor array, xv
mine shaft, 8
moisture content, 5, 66, 71, 119–120, 122,
  137, 154, 163, 213–214
Multi-Channel Analysis of Surface Waves
  (MASW), 149–151

N
neutron scattering, 66
neutron probe, 4, 66, 122, 135, 137
normally consolidated conditions
nuclear gauge, 66–67, 71, 119

O
oedometer test, 97
Ohm's law, 135
overburden drilling, 47
overconsolidation ratio, 97, 101, 108

P
P–S suspension sonde, 139
packer test, 81, 84–85

paleozoic, 17
passive earth pressure, 228
penetrometer, 116, 209, 211
penetration depth, 47, 81
percussion method, 47
permeability, 104, 188
physiography, 17
piezometers, types of, 185
    Casagrande open type, 192–195
    pneumatic, 188–192
    vibrating wire, 188–192
piezometric level, 14, 218
piezometric pressure, 46
piston sampler, 163–164
Pneumatic piezometer, 188–192
Poisson's ratio, 144
potential difference, 129–130, 132, 135
potential slip circle, 199
potentiometric elevation, 16, 19
precambrian, 17
pressuremeter, 50, 108, 121
probing, 65–68

Q
quaternary geology, 17

R
ray path, 128
Rayleigh and Love waves, 149
reclamation, 45–47, 121
re-compression index, 226
Reconnaissance Survey, 2
relative density, 3, 50, 71, 114, 121
remote sensing, 36
resistivity imaging system, 128,
    131–132
resistivity probe, 4, 66, 135
resistivity survey, 128–132
retaining structure, 38, 40, 45, 47, 199,
    224, 227–228
riser rod, 183
rising head test, 81
rock sampling, 157–158
rotary open hole drilling, 60–61

S
sampling
    groundwater, 158–160
    interval, 160
    planning for, 153
    rock, 157–158
    types of, 154–160
sample(s)
    disturbances, 176–180
    disturbed, 154
    extrusion, 176
    labeling of, 174
    preparation, 180–181
    preservation and packaging,
        171–174
    storage, 176
    transportation of, 175
    undisturbed, 154–157
samplers, 160–171
    large-diameter, 164–165
    Laval, 165–167
    piston, 163–164
    Sherbrooke, 167–171
    split spoon, 161
    U100, 162–163
    window, 162
sand replacement method, 119–120
satellite map, 8
Schlumberger configuration, 130–131
secondary compression index, 226
seepage, 3, 36, 38, 44, 54
seismic cone test (SCPT), 109–111
Seismic Dilatometer (SDMT),
    111–113
seismic reflection, 127
seismic refraction, 127
seismic signal, 146
seismic survey(-ing), 4, 127–128
seismogram, 146
seismometer, 127
self-boring pressuremeter test (SBPT),
    105–107, 126
    overconsolidation ratio, 108
    undrained shear strength, 107

settlement, 36, 38–39, 42, 44, 225
settlement markers/point, 184
settlement plate, 183–184, 199
shear modulus, 107, 109–111, 142, 144, 149, 151
shear velocity, 109, 114, 140
shear wave velocity, 5, 109, 111–112, 151
sheet pile, 47, 228
Sherbrooke sampler, 167–171
shore protection, 47
Singapore Marine Clay, 92, 94, 97–98, 101, 107
site reconnaissance survey, 23–24
slope stability analysis(-ses), 219, 221
spectrometer survey, 21
split spoon sampler, 161
spontaneous potential, 67
Spontaneous Potential (SP) probe, 4, 138–139
spot elevation, 8
soil
    cohesive, 50, 53, 57, 76–77, 154–156, 161–163, 180, 209–210, 212, 224–228
    granular, 3, 50, 53, 56, 67, 71, 109, 113–114, 120–121, 135, 154, 156, 180, 209–212, 224–228
    organic, 226
soil analysis
    dynamic, 143
    static, 142
soil sampling, 48
sonic logging, 139–142
Standard Penetration Test (SPT), 4, 69–75
stiffness, 225–227
strain path method, 98

strength parameters, 224–225
stress deformation analysis(-ses), 219, 221

T
transmission time, 128
transmission velocity, 127–128
transmissivity, 16
temporary dewatering, 38
test pits, 53–54
topographic map, 8–9, 16
topographic profile, 215
topography, 2, 23, 44, 48, 64
trenching, 54
triaxial test, 91, 96

U
U100 sampler, 162–163
unconfined aquifer, 14
unconsolidated undrained (UU) test, 224
undisturbed sample, 154–157
undrained shear strength, 90–97, 101, 107
Unified Soil Classification System (USCS), 209, 212–213, 215

V
vane equipment, 75
verification testing, 119–122
vibrating wire piezometer, 188–192

W
wash boring, 56
water displacement method, 119–120
water standpipe, 195–199
Wenner configuration, 130–131
window sampler, 162